KW-481-220

Heinemann Investigations in Biology

General Editor: S. M. Evans

Learning Genetics with Mice

In the same series:

Seashore and Sand Dunes
S. M. Evans and J. M. Hardy

The Behaviour of Birds, Mammals, and Fish
S. M. Evans

Units of Life: Cells and Unicellular Organisms
Richard Gliddon

Investigations in Woodland Ecology
C. T. Prime

Learning Genetics with Mice

Margaret E. Wallace, MA, PhD

Department of Genetics, University of Cambridge

AN LEABHARLANN,
WITHDRAWN
LEITIR CEANAINN.

**Heinemann Educational Books Ltd
London**

Heinemann Educational Books Ltd

LONDON   EDINBURGH   MELBOURNE   TORONTO   AUCKLAND
JOHANNESBURG   SINGAPORE   HONG KONG
IBADAN   NAIROBI   NEW DELHI

ISBN 0 435 60284 5

© Margaret E. Wallace 1971
First published 1971

Published by Heinemann Educational Books Ltd
48 Charles Street, London W1X 8AH

Printed in Great Britain by Richard Clay (The Chaucer Press) Ltd,
Bungay, Suffolk

Preface

This book describes a series of investigations in elementary genetics, using the laboratory mouse. This species is interesting because, for many genetic phenomena, comparison may be made with farm animals and man, with whom the student is familiar; yet it is attractive in its own right, easy to see and handle, requires no apparatus apart from cages, and breeds so prolifically that all investigations but one may be completed, if required, within a year. Many of the mutants described here are such as to stimulate thought about other biological investigations in which they may be used, such as anatomy, physiology, and behaviour.

The investigations are suitable both for project- and for class-work by sixth-formers, first- and second-year undergraduates, and students in colleges of education. Simple knowledge of chromosomal behaviour at meiosis and mitosis is assumed, and of Mendel's laws; but a glossary is provided for quick revision of terms used in elementary diploid genetics.

In twenty-five years' experience with this species for research and teaching purposes, the author has found a number of ways in which genetic segregations and their interpretation may be presented, which facilitate understanding by the novice as compared with conventional ways found in current textbooks. Full use of these is made in this work.

Current literature available at this level is somewhat out of date in methods described for the care of mice, and of genetic stocks in particular. This book describes modern labour-saving and hygienic methods of husbandry, as also modern methods of planning and recording during breeding programmes.

In many species used for the teaching of elementary genetics, insufficient attention is paid to the choice of mutant and to the residual genotype of the strains to be used. This sometimes results in so poor an agreement with expected ratios, due to impaired viability before classification or to imperfect penetrance, that the student in unconvinced; teachers are tempted to resort to providing demonstrations of 'good' data instead of allowing the student to provide his own data from real investigations. Realizing that disappointing results would very likely be obtained if stocks of mice available in this country and containing apparently suitable mutants were used in breeding programmes regardless of the full and residual genotype, the author gradually developed special stocks of mutants for teaching purposes. The mutants were chosen for their potential use in a number of investigations, and the stocks containing the different combinations of mutants were selected so as to maximize viability and to ensure easy and clear-cut classification. These stocks have largely been made generally available through Harris Biological Supplies; the investigations described in this book use a selection of them. Many students have done these investigations and shown them to be, under good husbandry, simply done, interesting, and fruitful.

Each investigation is based on published research. Much of it can be found in general works like *Biology of the Laboratory Mouse* (editor E. L. Green, McGraw-Hill, 1966) and *The Genetics of the Mouse* (H. Gruneberg, Bibliographia Genetica 15, 1952). These can be found in university libraries and from public libraries through the inter-library loans service. References to research papers not available through these two sources are given in the few cases necessary.

There are many more simple investigations in genetics which can conveniently be carried out with mice. It is hoped that this book will encourage the student to ask questions about phenomena related to those studied here – such as pleiotropy, penetrance and expression, quantitative inheritance and selection – and to be prepared to undertake the further investigations into these for which mouse stocks have been developed.

I am indebted to Mrs. Christine Finnegan of the Cambridge University Department of Genetics for technical assistance with the photographs in this book.

Cambridge 1971   M.E.W.

Contents

List of Plates

The solid piece of metal on the lid (*upper left*) is the shelter; when the lid is on the cage, the shelter lies over the nest area. The pelletted rat-cake is accessible through the sloping bars on the side opposite the nest (*upper right*). The open horizontal bars on the right lie, when the lid is on, over the exercise area (*opposite the nest*), where soiling occurs; so it is well ventilated.

The metal divider prevents rat-cake from falling into the bottle area. The bottle should not be filled fuller than shown or mice will not be able to obtain water. Its spout can be seen protruding through a hole on the right. The bottle cap rests against a stout horizontal bar which keeps it out of reach from gnawing teeth.

The piece cut out of the part of the lid facing the camera

accommodates the bulldog clip on the cage card when the lid is on. The paper-clip on the bottom of the cage card indicates that the mouse in the nest has a litter of six young.

The version of the Cambridge cage shown here is made by Associated Crates, and is available from them or from Harris Biological Supplies. The bowl shown is made of transparent polycarbonate, an asset for studying behaviour; it is also available in opaque polypropylene.

Plate 4   Cleaning the Cambridge cage   *facing page 21*
The hood, held in one hand over the nest, keeps the inmates undisturbed. The nozzle of a vacuum-cleaner, used in laboratories to remove soiled bedding, can be seen in position. In small breeding units, a spatula (*bottom left*) may be used to loosen the bedding, and the whole cage inverted, hood in position, over a suitable receptacle for the soiled material.

The cage-bowl shown is of opaque polypropylene. A similar version of the Cambridge cage is made by Cope & Cope; it has very rounded corners for easy cleaning.

Plate 5   Some albino locus phenotypes   *facing page 36*
The mouse in the foreground is extreme chinchilla, $c^e c^e$: fur coloured (pale fawn), eyes black. The mouse behind it is black-eyed white, $c^e c$: fur nearly white, eyes black. The mouse at the back and to the right is albino, cc: fur white and eyes pale pink.

Plate 6   The short-ear and normal phenotypes   *facing page 37*
The mouse on the left is short-ear, *sese*; compare with its long-eared companion. (The mouse on the right is normal at all loci; compare its plain sleek coat with those of tabby mice, Plate 8, and wavy-haired mice, Plate 7.) [Reproduced by permission of the *Journal of Biological Education.*]

Plate 7   Some mimic genes   *facing page 52*
Upper left: *wa-2wa-2 vtvt*. Lower right: *wewe Sd*+. Note that the waved-2 and wellhaarig genes each produces a waved coat and curled whiskers, and that vestigial-tail and Danforth's short-tail each reduces tail length. These mice are about twenty-eight days old.

The upper mouse is a tabby male, TaY. Note the bare patch behind its ears, the dark dorsal area in its agouti fur, and the bare tail. The fur looks greasy and the tail feels greasy; the knob near the tip of the tail is not present in all specimens.

The lower mouse is a tabby female, TaX. Note the stripes across her shoulders; facial stripes and the pronounced tail rings are not shown by all specimens.

1

AN LEABHARLANN,
CEARD CHOLÁISTE RÉGIÚN
LEITIR CEANAINN.

Recommended mutants

Wild-type colouring and colour mutants

Wild mice have a speckled greyish appearance. If an adult laboratory mouse of the same colour is held near the root of the tail and its hair blown forward, the base of the hairs may be seen to be black and the tips yellow. This pattern is called 'agouti' after the South American rodent of that name (Plate 1).

For a mouse to be of this wild-type phenotype it must carry the normal alleles of all mutants affecting colour, of which there are a great number. The better-known mutants are brown, b, albino, c, Maltese dilution, d, and pink-eyed dilution, p. A wild-type mouse must therefore carry B,C,D and P, as well as the normal alleles for all the less well-known mutants.

It must also carry A, the normal one of several mutant alleles now known for the A locus. The gene A is responsible for the yellow tips to the hair and has itself been called agouti. Each of the mutant alleles of A affects the amount and distribution of yellow pigment in the hair. Thus the allele non-agouti, a, produces no yellow at all. In an aa mouse carrying B,C,D,P, etc., the hairs are black the whole way along and the general colour is thus black (Plate 1). There are at least ten other alleles of A, but only two are useful for the present investigations. One is a^t, a gene which restricts yellow to the belly hairs; an $a^t a^t$ mouse (with B,C,D,P, etc.) thus has black all along the dorsal hairs but its ventral hairs are more yellow than black (Plate 1). It has been called black-and-tan, although the belly colour varies from tan to almost white. The other allele is A^y, which extends the yellow virtually to the full length of all the hairs. A mouse carrying A^y is therefore yellow all over the fur (Frontispiece).

The agouti locus is unusual, both in having several mutant alleles and in the dominance relations between them. However, as these features are useful in the following investigations, they are worth remembering. A^y is dominant to A, a^t, and a, so A^yA, A^ya^t, and A^ya mice are all yellow; A^yA^y on the other hand has no colour as it dies long before pigment is formed. a^t is dominant to a, so a^ta^t and a^ta mice are black-and-tan. But A is not fully dominant to a^t, so Aa^t mice have more yellow (or whitish-yellow) ventrally than AA, giving rise to the terms light-bellied agouti and dark-bellied agouti for their respective phenotypes.

Mutants at all other loci concerned with colour affect either the black pigment or the yellow, or both. The full phenotype thus depends not only on the mutant itself but on the agouti locus genotype. The brown gene, b, for example, changes black pigment to brown but has almost no effect on yellow. Mice of genotype $bbAA$, called brown agouti, thus have hairs brown at the base and yellow at the tip; bba^ta^t, called brown-and-tan, have dorsal fur brown and ventral fur whitish yellow; $bbaa$, called brown non-agouti, have fully brown fur (Frontispiece); and bbA^yA, called brown yellow, look deep yellow, there being no black pigment to be changed to brown.

In the last paragraph all the genotypes mentioned are assumed to carry C, D, and P; it is convenient to consider only one mutant at a time at the B, C, D, P loci, so the assumption made from here on is that for all unspecified loci the normal allele is present homozygously.

Maltese dilution, d, produces fewer pigment granules than its normal allele and clumps them, so that both yellow and non-yellow parts of the hair are pale. $ddAA$ mice, dilute agouti, are a washed-out bluish-yellow colour; $ddaa$, dilute non-agouti, a clear slatey-blue all over (Frontispiece). Pink-eyed dilution, p, however, affects non-yellow pigment more than the yellow. $ppAA$ mice look clear yellow, the non-yellow base of the hair being very pale, and $ppaa$ are a very pale bluish-grey colour (Frontispiece); in both genotypes the eyes are dark pink, the normal black pigment there being so pale that the red colour of the blood vessels shows and gives the eye its visible colour.

It is now apparent, from these descriptions, that the agouti gene A in combination with b, d, or p obscures somewhat the latter genes' effects on black pigment and that these effects produce the most attractive colours and the simplest to recognize. For this reason, most studies requiring classification at these and other colour loci are best done in mice homozygous for either a^t or a, the latter being the most favoured.

The best-known pink-eyed mouse is albino, cc. This gene allows no pigment of any kind to be formed, so that the coat is pure white. This locus, like the agouti, has several other alleles. These allow a little colour, but the one to be used in the following investigations is only slightly affected by agouti genotype. Called extreme chinchilla, $c^e c^e$ mice have dark eyes and are a very pale fawn colour in the nest, becoming darker, a pale coffee-with-milk colour, as they grow older. The c^e allele is not dominant to c, for $c^e c$ mice, called black-eyed whites, have dark eyes and a white (or very nearly white) coat. These genes c and c^e are usually best combined with aa for easiest classification of the various phenotypes (Plate 5).

Mutants not concerned with colour

There are mutants affecting every part of the body of the mouse which has been closely inspected. The following investigations use three hair-texture mutants and three skeletal ones.

There are four kinds of hair in the fur of a mouse, and their normal development can be altered in several ways. The gene Tabby, Ta, affects the distribution of oddly formed hairs in heterozygotes, giving a striped appearance. This is seen most clearly in agouti mice, AA, on the sides near shoulders and haunch (Plate 8). In some stocks the $Ta+$ phenotype is less clear than in others; so investigations with Ta should be carried out within the selected AA stock supplied. In $TaTa$ mice and males carrying the Ta gene, the phenotype is more abnormal, the coat appearing greasy and the area behind the ears almost bald (Plate 8). These mice also tend to have teeth which grow too long and require careful cutting at the tips; a mash diet sometimes helps the weaker ones to survive. The abnormal appearance of male tabbies is explained in Chapter 6.

The other two hair-texture mutants are waved-2, wa-2, and wellhaarig, we; wellhaarig is German for wavy. Mice homozygous for either of these mutants have curly whiskers, visible at about two days old, and wavy coats, seen most clearly at about fifteen days; in older mice the whiskers remain curly but the coat loses its waves and becomes rough and less shiny than the normal sleek coat (Plate 7).

Two of the skeletal mutants are also somewhat similar in phenotype. In both Danforth's short-tail, $Sd+$, and vestigial-tail mice, $vtvt$, the tail lacks some of the end vertebrae and is often bent; it varies in length (Plate 7). $SdSd$ mice have very short tails, and urogenital defects from which they die before birth or just afterwards. As with the Ta gene, it is advizable to breed only within the stock supplied, for this is selected to make the tails of viable short-tail mice long enough to serve as 'handles' for lifting and examination; mice with no tails at all are very difficult to catch.

The third skeletal mutant is short-ear, se. This gene affects the cartilagenous skeleton of many parts of the body, but these effects are too small to be seen easily. The clearest visible effect is on the ear. In normal mice at about eighteen days old, the pinna expands from a short fat pad to a large papery-thin organ; in $sese$ mice the ears look normal until eighteen days old, when they simply fail to expand, remaining short, fat, and crinkled (Plate 6).

Viability and classification

Normal wild mice are adapted to their environment by natural selection; mutants are therefore less well adapted. In laboratory conditions most mutants are at some disadvantage, though less so than in the wild; any defect lessens ability to survive, especially where there is competition with normal litter-mates. The next chapter will describe how to improve laboratory conditions so as to offset handicaps as far as possible; but even the best cannot compensate completely.

Of those mutants just described, the skeletal ones are the most likely to succumb under the slightest stress. Mutant mice which die before they are classified reduce the observed ratio of mutant to normal from that expected by a consideration of the gametic output of their parents – on which basis classical Mendelian ratios are determined. It is advizable therefore to classify all mutants at the earliest possible age.

It is also advizable to use mutants which can be classified at a young age rather than older, since this again obviates bias from those that die shortly after classification is possible. Many mutants can be classified from some juvenile feature long before the full phenotype is clear; thus wavy genes can be classified from whisker

Plate 1 Some agouti locus phenotypes
The mother is non-agouti, *aa*: black under her chin as well as over the rest of her body. On her back is a young black-and-tan, *aᵗa*: light on the belly, black dorsally. Behind them is a wild-type, agouti, *Aa*: its hairs are speckled. These young are about twelve days old.

Plate 2 Marking

Cutting a V-shaped piece out of the tip of the ear. The mouse is held firmly by the root of the tail, with slight tension against his four feet grasping the bars of the cage-lid. The scissors point from the head outwards, and the cut is made when a fold of the ear has been moved between the blades.

shape at two days whereas the coat feature appears a week or more later, and mutants giving a pink eye can be classified at birth whereas the coat colour is not fully developed for about two weeks. Where one colour mutant only is segregating, classification can often be made from the colour of the pigment under the skin, i.e. before hair is grown. Where more than one segregates it may be necessary to wait a little longer, but early classification may still be possible where the parts of the body affected are different for different colour mutants (e.g. for a^t,a the criterion is belly colour, and D,d or B,b can be discerned from dorsal colour only).

Table 1 summarizes the main features of all the mutants described in this chapter. It also shows the earliest age at which each mutant can be classified with certainty. However, with practice, and according to which combinations of mutants are segregating, certainty may be reached earlier and full advantage should be taken of this fact.

Fancy and popular names

Unusual mice have been cultivated for centuries, and many varieties have been bred by modern fanciers. Various names for the different phenotypes have become popular. Thus *bbaa* mice are commonly called 'chocolate', *bbAA* mice 'cinnamon', *ppaa* mice 'lilac', and *bbppaa* mice 'champagne'. These names were current before the genetics of the mouse had been fully studied. For some phenotypes there is more than one possible genotype; e.g. black-eyed whites can be genetically $c^e caa$ or a specially selected genotype involving two spotting mimics. It is therefore most unwise to assume that mice of a stated phenotype or fancy name have a particular genotype. Investigations should be started with mice of known *genotype*. A verbal description of a genotype, using the genetically accepted names for different mutants, as has been done for this chapter, is accurate but can be too lengthy, e.g. the fancy term a 'silver-champagne' would have to be called a pink-eyed brown dilute non-agouti. For this reason and for complete accuracy, symbols should be used in preference to words when ordering and recording mutants, and the conventions governing their use should be well understood.

Fancy names are intentionally omitted from Table 1.

B

TABLE 1

Recommended mutants and genotypes

Scientific name	Genetic symbol (linkage group)	Earliest age (days) to classify certainly	Some genotypes and their phenotypes	Genotypes available at Harris Biological Supplies	Numbers of the investigations using these genotypes
Yellow	A^y (5)	8	A^yA, A^yA^t, A^ya are all yellow, eyes dark. A 'sootiness', altering with the moult, is caused by other genes.	A^yA or A^yA^t	3
Agouti	A (5)	8	AA and Aa are dark-bellied agouti; Aa^t is light-bellied agouti. AA is wild-type.	AA	1, 2, 4, 5
Black-and-tan	a^t (5)	8	a^ta and a^ta are black above and yellow or whitish-yellow on the belly. Sometimes called tan-belly.	a^ta^t	1, 2, 4, 5
Non-agouti	a (5)	8	aa is black dorsally and ventral y.	aa	2, 4
Brown	b (8)	10	$bbaa$ is a uniform brown; bba^ta is brown above and whitish-yellow ventrally. $bbAA$ has hairs brown at base and yellow at tips.	$bbaa$	2, 4, 5
Albino	c (1)	0 (by eyes)	cc has a white coat and pink eyes; this is so whatever other colour mutants are present.	$ccaa$	1, 2, 5
Extreme chinchilla	c^e (1)	14	c^ec^e has a pale milky-coffee coat. c^ec has a nearly white coat. Both have dark eyes. Coat darkens with age.	c^ec^eaa	1, 2
Maltese dilution	d (2)	10	$ddaa$ is uniform slatey-blue; dda^ta^t is slatey-blue dorsally, whitish-yellow ventrally; $ddAA$ has hairs blue at base and pale yellow at tips.	$ddaa$	2, 4, 5, 9
Pink-eyed dilution	p (1)	0 (by eyes)	$ppaa$ is a uniform pale bluish-grey; ppa^ta^t is pale bluish-grey dorsally, whitish-yellow ventrally; $ppAA$ is a clear sancy-yellow. All have deep pink eyes.	$ppAA$ or $ppaa$	2, 4

Name	Symbol			Description	Genotype	References
Danforth's short tail	Sd	(5)	0	Sd+ have shorter tails than normal; variable. SdSd have no tail, kidney disorder, and die perinatally.	Sd+ bbaa	3
Short-ear	se	(2)	18	sese have short ears, and sometimes hydronephrosis. +se have normal ears and rarely hydronephrosis.	sese AA and +se AA or sese aa and +se aa	2, 4, 8 2, 4, 8, 9
Tabby	Ta	(20)	14	TaX is a female with faint dorso-ventral stripes on sides; TaTa and TaY look greasy, have a bald patch behind ears and faulty teeth; TaTa is female, TaY male.	TaX AA or TaY AA	7
Vestigial tail	vt	(7)	0	vtvt are similar to Sd+; variable.	vtvt (wa-2, and colour mutants may appear also).	10
waved-2	wa-2	(7)	2 (whiskers)	wa-2wa-2 has waved hair and curly whiskers, the coat becoming rough with age.	wa-2wa-2(vt and colour mutants may appear also).	6, 10
Wellhaarig	we	(5)	2 (whiskers)	wewe is closely similar to wa-2wa-2.	weweaa (b may appear also)	6, 11 12
				For linkage programmes:	++/sed, aa and sed/sed, aa	8, 9
					$\dfrac{a^t + Sd}{a\ we\ +}$ and $\dfrac{a\ we\ +}{a\ we\ +}$	11, 12
					each with +b or bb	11, 12

Symbols and genotype conventions

The use of a capital letter for a dominant and a small letter for a recessive is well known. Thus the moment a symbol for a mutant, e.g. d, is used, one knows that it is a recessive one. A second letter is added to the first where the single letter has already been ascribed to another mutant: thus se was given to short-ear because s had already been given to recessive pied. Alternative genes that can occupy the same locus are called 'alleles', and where more than two are known, superfixes are used to distinguish them (e.g. c and c^e); figures are occasionally used, e.g. wa-2 is so-called because a similar mutant was already known.

In some textbooks an oblique line is used to separate the two alleles of a genotype, e.g. D/d for heterozygous dilute. However, particularly among mouse geneticists, this convention is giving way to one whereby the oblique is only used for linked genes (see below). Thus the latter genotype is more usually written Dd; even this is less clear than is often needed. Thus Dd indicates a heterozygote but, without other information, it does not indicate which is the mutant, D or d. A most useful convention is to use a symbol for the mutant and $+$ for the normal. Then $+d$ and $Sd+$ are clearly both heterozygotes, the former for a recessive mutant and the latter for a dominant one. (Note that the dominant allele is written first in both cases.) If the genotype is being written along a line, as here, the $+$ refers to the symbol closest to it, a comma being inserted as necessary; thus $+d+b$ and $Sd++b$ are double heterozygotes which would otherwise have been written $DdBb$ and $SdsdBb$. It is conventional to maintain the same order for the details of two loci, e.g. the mating of homozygotes unlike for d and b is written $++dd \times bb++$; here those for the brown are written first then those for the dilute. Where there is any chance of doubt, genotypes may be written one above the other, when details for one locus occupy one column, details for the next a second column, and so on, e.g.

$$\left. \begin{array}{l} +\ +\ d\ \ d\ a\ a \\ b\ \ b\ +\ +\ AA \end{array} \right\} \text{ for } BBddaa \times bbDDAA.$$

In the case of linked genes, it is conventional to write their symbols in the order in which they are situated on the chromo-some, e.g. a^t *we Sd*. If only two are concerned, they are written in order of most common usage, e.g. *sed*. Where there are various possible arrangements, clarity is again achieved best by writing the alleles on one chromosome above those on the other, e.g.

$$\frac{a^t + Sd.}{a \; we \; +}$$

Here the $+$ on the top line means normal for the *we* locus, and that on the lower line means normal for the *Sd* locus. If the line dividing the two chromosomal genotypes is slanted to the oblique, this arrangement may be written on one line:

$$a^t + Sd/a \; we \; +.$$

This is more convenient but less clear.

The conventions used throughout the following chapters are those in common use by mouse geneticists (Green, 1966). If the reader is used to different conventions, he may avoid confusion by writing the genotypes in his usual manner, with the unfamiliar convention alongside, until the latter becomes familiar.

Reference

Green, E. L. (1966). Chapters 6 and 8 in *Biology of the Laboratory Mouse.* 2nd edition. New York: McGraw-Hill.

2

Care of genetic stocks

Textbooks vary in their advice on the subjects covered in this chapter. The recommendations made here are based on twenty-five years' experience. References are given at the end of the chapter for sources of cages, ancillary equipment, and visual aids.

Life-cycle and management

The mouse is useful for genetic work because, unlike man and mammals of economic importance, of which it is a prototype, it breeds prolifically, has short generations, and is cheap to house.

A thorough knowledge of the life-cycle is necessary for good management, and this in turn is necessary if the full potential of the species for genetic work is to be realized. Details of the cycle vary with the stock or strain, inbred and mutant mice in general breeding later, less long, and less prolifically than outbred. The following rough guide is summarized in Table 2.

(a) Development
Newborn mice, naked and bright pink, become paler in a few hours; as they suckle, milk becomes visible in the stomach. The sexes are distinguishable at birth, the penis being slightly larger than the vulva, and the distance between these organs and the anus being greater in the male; this difference in distance becomes more marked until the growth of fur obscures it.

The colour of the eyes is apparent at birth and for a few days afterwards; it can be seen again when the eyes open. At about 2 days old the whiskers have grown enough to be visible, and in a further day or so the ears open. At 5 days, pigment can be seen in the skin, and hair is half-grown by 10 days; the agouti phenotype can then be discerned, for the yellow hair-tips look like yellow dust over the darker pigment. At 8 to 10 days the teats are visible in females, a feature which may be used as a sexing check.

TABLE 2
The main features of the life-cycle

A Development

0–1 days: Naked, ears closed, whiskers very short, eye-colour visible through closed lids, sex just discernible.

2–3 days: Ears open, whisker shape visible.

5–6 days: Darker pigments visible in skin.

8–10 days: Hair half-grown and colour visible. Teats visible.

12–14 days: Eyes open, teeth erupting, weaning and first moult beginning, fatty layer reduced, face elongating. Greater activity sometimes accompanied by slower weight gain.

18 days: Outer ear expands.

21–23 days: Weaning complete, mice very active and appearance adult. Sexing hindered by fur.

35–46 days: Vagina open, descended testes visible (but retractable); sexing clear. Weight about 20 g, mutants usually less.

B Reproduction

Sexual maturity: 6 weeks usually, later in some inbreds and mutants, earlier in some strains.

Oestrous cycle: 4–5 days. Vaginal plug indicates copulation within last 24 hours.

Gestation: 19–21 days; up to 35 if lactating.

C Breeding performance

Generation interval: Shortest $2\frac{1}{4}$ months, longest about 18 months.

Breeding span: Females up to 6–12 months old, males up to $1\frac{1}{2}$ years.

Life-span: Females up to 1–2 years, males up to 2–3 years.

Litter interval: About one litter per month if male always present.

Output per female: Variable, depending on genotype. 20–120. Rough expectation for good breeding stock: 1 baby mouse per female per week.

At 12 days the eyes open, and the teeth begin to erupt in a further day or so.

Weaning begins at about 14 days, when eye and tooth development enables food and water to be sought and consumed. At the same time, the mice quickly become more active, the head becomes more pointed and the general shape slims as the insulating fatty layer disappears with the first moult.

At about 18 days, the pinna of the ear expands rapidly. At 21 to 23 days weaning is complete and the mice are at the most active stage of their lives. Gentle handling of the cage and lid, and slow confident handling of the mice themselves is needed to avoid escapes.

From 23 days on, those wanted for breeding may be separated from their parents; they must be moved before 30 days to avoid unplanned mating with parents. Unwanted littermates can be killed when classification is complete, certainly before their mother's next litter is born.

At 30 to 35 days the vagina opens; at 35 to 40 days the testes descend.

(b) Reproduction

Before sexual maturity at 5 to 6 weeks, the two sexes must be housed separately or with their assigned mates. Sexually mature males are aggressive to strangers, so smaller ones should not be housed with larger ones. If there is fighting when housing together equally-sized strange males, remove them all to a clean cage with plenty of wood-wool to hide in.

Caging pairs together before maturity does not impair breeding, but in a cool room caging in single sexes in larger numbers (3 to 6) ensures warmth; they will then be healthier and may therefore breed better when eventually caged for breeding in pairs and trios. However, it is inadvizable to leave mice caged in single sexes for more than two months, especially if overcrowded, as health and fertility suffer.

The oestrous cycle takes 4 to 5 days. Copulation, usually at night, is followed by the formation of a white vaginal plug; if this is seen, by simple external examination, one may be sure that if conception has occurred, it has done so within the last 24 hours. In a warm even temperature breeding occurs throughout the year. If the male is present at parturition, conception occurs almost immediately, and litters may be expected at the rate of about one per month. If young are disposed of early, lactation is curtailed, and the gestation period, which during lactation may be as long as 35 days, is reduced almost to the 19 to 21 days usual for non-lactating females. Clearly if the female is to be remated, her original male must be removed before parturition.

The simplest mating system is permanent monogamous pairs. Trio mating (two females to one male) is economical of food and space, and is feasible if young are disposed of soon after birth. But if there is likely to be doubt as to which female is the mother when two litters are born together, or if the smaller litter, or small mutants of either litter, are likely to suffer from competition for milk, it is better to separate the first female of a trio seen to be pregnant. To avoid fighting, she should not be returned to the remaining pair until two days after her litter and that of the other female are disposed of.

The expectant and nursing female and young can be gently handled daily, but the nest must not be pulled out of shape and the young must be kept warm in one's hand. If the nest is unduly disturbed in the week around birth, the young may be abandoned and eaten. To give plenty of time for nest-building and settling down, a female which must be moved to another cage should be moved as soon as her outline indicates pregnancy; this is usually about a week before the litter is due.

(c) Breeding performance
Females breed until 6 to 12 months old, males up to 18 months. With 4 to 9 young to a litter, or as many as 15 in crossbred mice, the total progeny of a female may number between 20 and 120. A figure useful in planning is as follows: for pairs and trios where the male is present all the time, and the females killed at 8 months old, a reasonable expectation is about one baby mouse per week per female. With a generation interval of about 3 months, an average expectation is one generation per academic term. With care, all the following investigations (except the full 5 generations for *wa-vt* in Chapter 6) may be completed within a year.

Handling, marking, and killing

(a) Handling
Frequent inspection and gentle handling ensure quiet mice and good breeding performance. If nestling mice are inspected separately, they must be returned within seconds and in a warm hand to the nest. After about ten days nestling mice should always be picked up by the tail, near the base. Once out of the nest they feel more secure if, while the tail is held, their forefeet are allowed to grasp the cage-lid or one's sleeve; this allows examination of

the belly and hind-feet. For dorsal inspection, their hind-feet should be allowed to rest; slight tension on the tail inhibits restlessness and prevents the mice running up their own tails. Tail-less mice may be inspected easily if they are picked up by the loose neck skin and put gently in a transparent ventilated box.

(b) Marking

For the following investigations, the sex and phenotype of mice need not be distinguished before they are 18 days old, and usually need to be distinguished after this age only when mice of the same phenotype from different litters are stored in single sexes, or when similar females are mated in a trio. Once the pinnae have expanded, a V-shaped piece may be cut out of one ear painlessly with sharp scissors. As skill is required to obtain a small clean cut visible in dorsal view, especially on short-ear (*sese*) mice, a dead one should be practised on first. Place the living mouse on a cage-lid, with all its feet grasping the bars, and the tail held with slight backward tension; move the ear with the scissors so that a fold forms between the blades, and snip (Plate 2). The left ear is easiest for right-handed people. If more possibilities are required than single clips for either or both ears, the last 5 mm of the tail may also be cut off; this gives eight possibilities altogether.

(c) Killing

Most laboratories use toxic gases to kill mice. Danger to people, as well as humane treatment of animals, must be remembered at all times. Use as small a killing jar as the numbers normally to be killed will allow; a transparent jar enables one to see that enough agent is used to kill quickly. The gas dispenser should also be small, and both the dispenser and the killing jar must have tight-fitting lids. For ether, an unbreakable plastic hair-spray bottle is suitable, as the jet aperture is too small to allow dangerous amounts of vapour to escape and it ensures control of the quantity of liquid poured into the jar; the latter should be lined with absorbent material and the dispenser directed on to this and not on to live animals. A notice concerning inflammability should be fixed to ether and chloroform containers. Carbon-dioxide, coal-gas, and natural gas can also be used, with similar precautions.

Control of disease

Obviously diseased animals should be killed for humane reasons. If adult mice are found dead in a cage, the healthy occupants should be given a clean cage and the first cage should be washed thoroughly. Cages containing slightly sick-looking animals – they sit in a hunched position and the coat loses lustre – should be identified (see next chapter) and inspected daily until recovery or certain diagnosis. Early diagnosis and culling may prevent ulti-mate sacrifice of the whole stock.

Most mice carry a number of pathogens without obvious symptoms. Disease then becomes overt because the mice have been stressed in some way: overcrowding, temperature fluctuation, prolonged travel, or unsuitable caging, can each precipitate disease. The commonest endemic diseases are: infantile diarrhoea (seen as a yellow liquid in young 6–12 days old, which may stick to their coats), pneumonia (there are many pleuropneumonia-like organisms with various respiratory symptoms), and mouse catarrh. Some of these, especially the first, can be shaken off in a generation by fostering newborn mice with healthy mothers, but for most of them improved caging and hygiene are necessary to make the disease 'go underground' again, as in normal healthy mice. Catarrh, suspected if mice make frequent 'schik' sounds as they move about, may more certainly be diagnosed if these noises continue when the mouse is dangled close to one's ear. As it can cause sterility before the affected mice look sick, one needs to be on the alert for this disease and to cull as early as possible.

Most diseases are difficult or expensive to cure, and this requires veterinary knowledge. If they become widespread, the cheapest procedure is to kill the whole stock, disinfect all equipment and start again with healthy stock from a supplier. Prevention by the provision of a good environment (see the next section) is thus a far better goal than cure.

Caging, cleaning, bedding, and diet

(a) The room
The room in which the cages are to be kept should not contain other rodent species, or be concerned with activities likely to

introduce noxious gases or sudden loud noises; it should have an even temperature (18–20°C is ideal), good ventilation (eight changes of air an hour if possible), good natural light (or artificial light controlled by a time-switch), and a humidity close to what is comfortable for man.

Demonstration boxes should not be left in strong sun, on a radiator, on the floor, or near a fan or open window. There should be sufficient permanent wall-racking for the whole colony.

(b) The cage

The author has designed a cage which gives exceptionally good breeding results and a lower servicing time than any other (the 'Cambridge' cage, Wallace 1965, 1969). Its special features are given briefly below; these should be given attention where the choice available excludes the Cambridge cage.

When the cage is correctly assembled (Plate 3), there is a nest area (under the shelter) and an exercise area (on the opposite side). Food and water are accessible only in the latter area. This design allows mice to shore bedding up to the lower edge of the shelter, so making the nest area like a tunnel opening towards the bottle end. This enables them to keep nest temperature high for newborn young (much higher than room temperature), and to lower it gradually, as the young grow hair, by opening out the nest and then removing shored-up bedding. It also ensures that they reach food and water by crawling under the bottle, so keeping bedding away from the spout; this prevents the bottle siphoning out and flooding the cage. Mice urinate and defaecate while moving about in the exercise area, and the design, with open bars on this side, ensures good ventilation of waste matter so that it dries up and does not smell unduly or harbour pathogens.

Damp bedding is the bane of most mouse-breeders. Apart from the above design features, and the specially narrow bore of the bottle-spout, extra care by the user of this cage can obviate dampness as follows: press the cap firmly on to a newly-filled bottle and allow the bottle to drip outside the cage before inserting it; and put the bottle-spout up, in the cage, if it is to be carried far.

(c) Cleaning

Unless the Cambridge cage is overcrowded, it needs cleaning only once in every one or two weeks. Weanlings, stored mice, and matings not actually breeding, should be transferred to a clean bowl with fresh bedding, and the lid and bottle of the dirty cage transferred to the clean one. Food and water may be replenished at the same time; bottles should not be filled higher than 2 cm below the cap when standing. Pregnant or nursing females need not be so disturbed: enclose the nest and inmates in a hood (made by quartering a spare cage bowl), loosen the soiled bedding with a stout plastic cooking spatula, and turn the cage upside-down over a suitable recepticle (Plate 4); return it to the right way up, replenish bedding, remove the hood and return the lid. This procedure prevents many litter-losses otherwise suffered after the nest or young are disturbed by transfer.

The bottle can be cleaned with a stout bottle-brush, the cap with a broad paint-brush, and the tube (left in the cap), with a 2 mm diameter tube brush (catalogue no. BU 580, Gallenkamp, Technico House, Christopher Street, London E.C.2).

All cage parts require hand-hot water, mild detergent, and thorough rinsing.

(d) Bedding and diet

The most hygienic and cheapest materials are sawdust and fine quality wood-wool. The sawdust should be spread *thinly* over the cage bottom, and the wood-wool twisted into a rough nest in the nest area (Plate 3).

Pelletted diets have replaced all others because they are nutritionally balanced, cheap and hygienic. Local corn merchants and pet shops keep a rat-cake suitable for mice; or a local laboratory housing rats or mice will usually advise on supply.

Ordering mice and cages

Mice and cages should be ordered at least a month ahead of requirements. Harris Biological Supplies Ltd., Oldmixon, Weston-super-Mare, Somerset, stock the Cambridge cage and all the genotypes needed for the following investigations. Mice should be ordered with reference to genotype and type of investigation, with a second choice in case of short supply.

At present only a few of the genotypes described here are available with other supply firms. Application for mice from supply firms or laboratories should emphasize that guaranteed genotypes are required. Few schools keep mice and records, as yet, in such a way as to guarantee genotype, although this situation may improve.

If mice are to be sent anywhere, they should be packed in containers like those used by suppliers, with rat-cake or wheat as food, and cut potato and carrot as a source of water. They should not be sent in cold weather, nor on journeys likely to keep them out of proper cages for more than sixteen hours.

References for further reading, and sources of cages, ancillary equipment, and visual aids

Wallace, M. E. (1965), 'The Cambridge mouse cage', *J.A.T.A.* (now *J.I.A.T.*) 16, 48–52.
  The Cambridge cage may be obtained from Cope & Cope Ltd., 57 Vastern Road, Reading (Reading 54491/2); from Associated Crates Ltd., Coronation Avenue, Stockport (Stockport 3016); and from Harris Biological Supplies Ltd., Oldmixon, Weston-super-Mare, Somerset (Weston 27534).

Wallace, M. E. and Hudson, C. H. (1969). 'Breeding and handling of small wild rodents: a method study'. *Laboratory Animals*, 3, 107–17.

Wallace, M. E. (1968). 'A chute for the transference of hyperactive mice during cage-cleaning procedures'. *Laboratory Animal Care*, 18, 200–5.
  This chute can be made in cardboard from the instructions in the above reference.

Wallace, M. E., Gibson, J. B., and Kelly, P. J. (1968). 'Teaching Genetics: The Practical Problems of Breeding Investigations'. *J. Biol. Educ.*, 2, 273–303.
  Copies of this paper are obtainable from Harris Biological Supplies (see above).

Wallace, M. E. 'Development of normal mice'. 4 min. film loop in colour, with guide notes. 8 mm. Macmillan & Co. Ltd., Little Essex Street, London W.C.2.

Wallace, M. E. 'Handling and management of normal mice'. 4 min. film loop in colour, with guide notes. 8 mm. Macmillan & Co. Ltd., Little Essex Street, London W.C.2.
  This loop also shows assembly and cleaning of the Cambridge cage.

3

Planning and Recording

Choice of an investigation

(a) Design

The choice of an investigation should be based primarily on what is to be discovered or demonstrated. Seldom can more than one genetic principle at a time be convincingly elucidated. The beginner is unwise to embark on an investigation which may produce complex data from which no firm conclusions can be made. A common 'experiment' is to take two very different-looking mice of unknown genotype and mate them together to 'see what happens' – with no idea of what might happen. Usually the result is a multitude of phenotypes in quite unrecognizable ratios, for two or more of the following phenomena have been involved simultaneously – semi-dominance, epistacy, linkage, sex limitation, etc.

The most important factor in the success of an investigation is undoubtedly simplicity of design and object; the remaining chapters of this book are written with this in mind.

(b) Cost

In most institutions time and money are limited, and the choice of investigation cannot finally be made until several desirable ones have been compared for these aspects.

The initial outlay of an investigation is easy to assess, but this cost is usually less than that of running it. To assess the latter requires the estimation of numbers to be bred in each generation, how long each mouse must live before it can be disposed of, and the cost of diet, bedding, and of some labour, e.g. for servicing and simple management. A rough guide from laboratory practice is as follows: each mouse maintained beyond a few days from birth costs between 0.5p and 1.5p a day. Thus an investigation requiring 100 one-month old progeny bred in the generation

giving data on ratios, and some 20 mice in the preceding generations, each living for at least 3 months, involves 160 months of
mouse-maintenance time. At 1p a day, this comes to £50. The
initial trio, on the other hand, might cost a good deal less than
£5; and if no more than 12 cages were required at any one time,
these at £1.50 each would cost £18. Thus the initial outlay for
mice and cages is less than £25, against £50 for running costs;
and the latter rise with each mouse kept longer than necessary.

Clearly, therefore, it pays to plan carefully and to cull unwanted mice regularly (see below), and those investigations
needing the larger progenies will, in the main, have higher running costs.

A small saving in cost may be made by starting with genotypes
which can be used in several investigations. The last column of
Table 1 (Chapter 1) shows that nearly all are usable in two or
even more investigations.

(c) Holidays

The feasibility of an investigation may also rest to some extent
on what can be done with the mice out of term. Where no
permanent technical assistance is employed, as in many schools,
mice can be given to students to look after in their homes; but
this works well only if the students understand the investigation,
keep up a regular routine of servicing and recording, and keep
the mice in a warm, draught-free room.

Most of the investigations described in this book can be planned
to finish before the long summer break. Then only those mice
need be kept which are required to maintain essential genotypes.
However, disposal of all stocks and re-ordering from the supply
firm for the new academic year, is often less expensive; and it does
give complete confidence in the genotypes of the mice to be used.
Some schools in a district arrange that, during holidays, each
takes on one organism and maintains it for the others.

Planning an investigation

Investigations with mice cannot be carried out in a month; for
this reason they need both long-term and short-term planning,
i.e. an over-all programme, and a regular routine to carry it out.

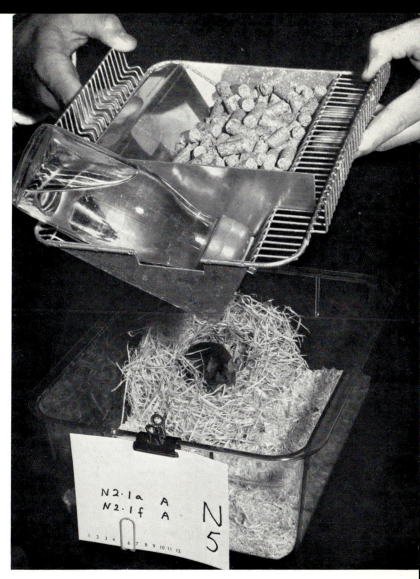

Plate 3 The Cambridge cage

The solid piece of metal on the lid (*upper left*) is the shelter; when the lid is on the cage, the shelter lies over the nest area. The pelletted rat-cake is accessible through the sloping bars on the side opposite the next (*upper right*). The open horizontal bars on the right lie, when the lid is on, over the exercise area (*opposite the nest*), where soiling occurs; so it is well ventilated.

The metal divider prevents rat-cake from falling into the bottle area. The bottle should not be filled fuller than shown or mice will not be able to obtain water. Its spout can be seen protruding through a hole on the right. The bottle cap rests against a stout horizontal bar which keeps it out of reach from gnawing teeth.

The piece cut out of the part of the lid facing the camera accommodates the bulldog clip on the cage card when the lid is on. The paper-clip on the bottom of the cage card indicates that the mouse in the nest has a litter of six young.

Plate 4 Cleaning the Cambridge cage

The hood, held in one hand over the nest, keeps the inmates undisturbed. The nozzle of a vacuum-cleaner, used in laboratories to remove soiled bedding, can be seen in position. In small breeding units, a spatula (*bottom left*) may be used to loosen the bedding, and the whole cage inverted, hood in position, over a suitable receptacle for the soiled material.

The cage-bowl shown is of opaque polypropylene. A similar version of the Cambridge cage is made by Cope & Cope; it has very rounded corners for easy cleaning.

(*a*) *Long-term*

Long-term planning involves the estimation of numbers of mice to be bred in each generation and, from this, the number of cages needed. Most investigations need the biggest progeny in the generation producing the most complex ratio. The size of this progeny determines how many matings are required to produce it, and thence the size of the antecedent generation. Thus the best way to make these estimates is to start at this critical generation and work back; only then can a realistic number of mice be ordered for the first-generation matings.

When ordering cages it is advisable to estimate the number needed for the time when the biggest progeny is to be bred; it will become clear in the earlier generations if this is an underestimate, and there will still be time to order more.

Recommended numbers of matings and sizes of progeny are given in Table 3 for the investigations described in this book. However, the student who works these out himself will begin to understand more thoroughly the underlying genetics of his programme. He will also see that a knowledge of simple statistics ensures that estimates of numbers to be bred allows the expected ratios to be recognized despite chance deviations. A simple way to allow for unforeseeable accidents is to double the estimated number in Table 3 of breeding females for the critical generation.

In this critical generation there is a choice: to breed all the progeny in the shortest possible time, i.e. from first litters only, or to breed them over a longer time, i.e. from several successive litters. The former saves time, and the latter saves space since fewer matings are needed. Where first litters only are to be used the number of matings needed depends on the average size of first litters: if this is 5, and 100 progeny are to be bred, the number of matings needed is $100 \div 5 = 20$. Where successive litters are to be bred in, say, one academic term of $12\frac{1}{2}$ weeks, and 100 progeny are required, a rate of one baby mouse per week per female (Chapter 2) may be assumed; then the number of breeding females needed is $100 \div 12\frac{1}{2} = 8$.

(*b*) *Short-term*

The short-term routine is simple. Whatever the investigation, the inmates of every cage should be examined regularly once a week,

c

TABLE 3

Recommended numbers of matings and Progeny for each investigation

Investigation	(Chapter describing it)	Parental mating Gen. 1	Gen. 1: No. of mated females	Gen. 2: No. of mated females	No. of progeny from Gen. 2
*Two alleles with no dominance	(4)	$c^e c^e \times cc$	3	4	32
*Two alleles with dominance	(4)	$BB \times bb$	3	4	32
*Two alleles with dominance and lethality	(4)	$Sd+ \times Sd+$	3(12)†	5(15)†	32(120)†
*Two pairs of loci: with dominance	(5)	$pp++ \times ++sese$	4	6	46
*Two pairs of loci: epistacy	(5)	$cc++ \times ++dd$	4	6	46
Two mimic genes	(5)	$++wawa \times wewe++$	3	4	30–50
Sex linkage	(6)	$TaX \times XY$	3	4	32
Close autosomal linkage: repulsion	(6)	$++dd \times sese++$	3	4	32
coupling		$++++ \times sesedd$	3	4	32
Loose autosomal linkage:	(6)	$wawavtvt \times ++++$	6	8	100–200
Three linked loci:	(7)	$\dfrac{a^t + Sd}{a\ we +} \times \dfrac{a\ we -}{a\ we -}$	10‡	—	200‡
Three linked loci and one independent one:	(7)	$\dfrac{a^t + Sd\ B}{a\ we + b} \times \dfrac{a\ we + b}{a\ we + b}$	10‡	—	200‡

* Other genotypes are described for these investigations in the relevant chapter.

† Bracketed numbers are a Gen. 2 progeny big enough to discern an expected 2 : 1 from 3 : 1.

‡ Numbers assume all progeny are from one backcross generation. If Gen. 2 mates are bred from Gen. 1, and all matings breed to infertility, 3 to 4 pairs are enough for Gen. 1.

records made of their progress and any young classified as far as possible. If records are made less often, a number of things may go wrong, e.g. it may be impossible to ascribe a new litter to the correct female of a trio – and this is important if one female differs from the other either in fertility or genetically. With weekly examination and culling of some fully classified young, in-litter competition can be reduced and the vigour of those mice needed for the next generation enhanced.

With good cages and an efficient cleaning and recording system, all but the most complex experiments take not more than *one hour a week for four cages*, less with experience. If the student knows at any one time exactly how many matings he still requires and how many young for for the next generation, he can also keep the number of cages to a minimum; for as the new generation becomes available most of the parents can be disposed of. It is advizable, however, to keep one mating of the old generation breeding as a safety measure until all the new generation have begun to breed.

Short-term efficiency reduces time, energy, and cost in the long run.

Record system

All breeding records should be brief and clear. For brevity, symbols can be used instead of words, and much-used words like 'date of birth', 'number in litter', and 'phenotype' can be omitted. For clarity, a key to symbols should be readily available, as also a key to a typical page, so that information relevant to such omitted words is recognized by position.

Permanent and temporary records should be kept separately. Those described below have stood the test of ten years' use in a large laboratory, being simple enough for the simplest experiments, yet flexible enough for a great variety of experiments including complex ones.

Permanent records should be made on long-lasting materials, and entries rigorously standardized, so that students of one year may use those made, say, ten years ago. The reference system should allow an individual mouse's ancestry to be traced quickly, and data from similar matings to be removed and considered as a whole for analysis. Interleaved blank pages in students' copies of permanent records allow personal notes to be made. Temporary records should contain current information, i.e. concerning the present inmates of cages and their present progress.

(a) Permanent record system

A loose-leaf hard-cover file is required, with stout paper lightly printed in $\frac{1}{4}$ inch (or 0.5 cm) squares. If two or three investigations are to be recorded in one file, dividers of tough card may be used to separate records for different investigations. All these are obtainable from: Mr. G. A. de Vall, Waterlow & Sons Ltd., 85–6 London Wall, London E.C.2. Specifications: 'black hard-cover ring-binders (3 rings); loose-leaf sheets $9\frac{1}{2}$ in. $\times$ $7\frac{1}{2}$ in. ruled quadrille both sides and punched three holes and creased; and coloured card dividers'. Figure 1 shows a sample page in the permanent record file. It contains the mating and litter record for one pair of mice.

(i) *Mating record* Each mating is recorded on a separate loose-leaf page. All matings involved in one investigation are given the same reference letter, e.g. M for a study of monohybrid inheritance. This letter is written at the top left side of the page.

The pages are numbered serially, the number being written after the reference letter: a new mating is given the next number available, e.g. the second mating has M2 at the top left side of the page. All pages are kept in serial order, but can be removed temporarily and rearranged for collation of data. Terminated mating records are removed to a separate loose-leaf file.

For each mating the following details are recorded. The date when male and female were caged together is written at the top right side of the page. The reference number, genotype, and date of birth of the female is written towards the top across the middle of the page, and similar details for the male are written directly underneath.

(ii) *Litter record* Details of each litter born to this mating are given in the space below the mating details. When a new litter is first seen, the following record is made, all on one line: first the litter number (first, second, third, etc. to this pair), then the number of individuals in the litter, and then the date of birth

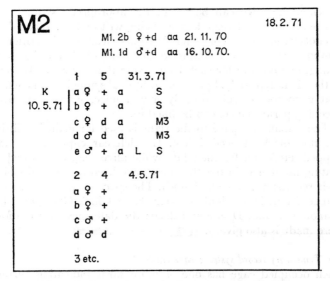

Figure 1. *A sample page in the permanent record file*

Key to lay-out:

M2 = mating number.

18.2.71 = date when female and male were caged together.

M1.2b = reference number of the female: M1 is the mating number of her parents; they were the first pair to be mated in this investigation; 2 shows that she was a member of the second litter of mating M1; b shows that she was member b of that litter.

M1.1d = reference number of the male; he was member d of the first litter of the pair forming mating M1.

21.11.70 and 16.10.70 are the dates of birth respectively of the pair forming mating M2.

Line reading '1 5 31.3.71' means that the first litter consisted of 5 mice born on 31.3.71.

L = left ear-clip; S = stored; M3 shows that members c and d have been mated as mating M3. K = killed.

(estimated from external features and card clip information, see below). Each member of the litter is given a letter, and these letters are written one below the other; in this way each member has a blank line on which its full phenotype will gradually be written.

When the mice can be sexed, the appropriate symbols are written beside the letters, females being written first. When their phenotype for other characters becomes clear, this is written in symbol form in the columns to the right of that assigned to sexing, one column for each locus. Note that the symbols apply to the phenotype only. Space for additional symbols can be left so that genotypes of subsequently mated mice, when clear from breeding performance, can be filled in.

The remaining space to the right is used for artificial distinguishing marks if needed (e.g. L = left ear clipped, T = tail clipped, etc.) and for their fate while alive (e.g. S = stored; a mating number indicates that they have been mated and where their records may now be found). The space on the extreme left is used for their terminal fate (e.g. K = killed, G = gone, i.e. cannot be found, D = found dead; the date these observations were made is also given).

(b) *Temporary record system: cage cards*

Each occupied cage has one or more cards indicating current information as to its contents. Those cards recommended for use with the Cambridge cage are obtainable from Mr. K. Brett, Galloway & Porter, 30 Sidney Street, Cambridge. Specification: 'cards $4\frac{3}{4}$ in. $\times$ $3\frac{3}{4}$ in. quadrille ruled on one side only, with numbers 1 to 12 printed on the bottom left line of squares'.

When a mating is recorded, a card is filled in for it. The card shows only the mating number, the reference number of the female and male, and their phenotype and artificial markings, if any. It is attached to the cage the mated pair occupies (Plate 3). In trio mating, two such cards are put on the cage, one for each female. If one female is transferred when pregnant to another cage, her card is put on that cage, the remaining card being left with the other female and the male.

A card is destroyed when the mating is terminated; this can be when one or both members of the mating dies or is remated. The lay-out of the information of these cards must be exactly as shown (Figures 2a and b).

Week-by-week changes in the contents of a cage can be indicated by the use of coloured plastic paper-clips. These are obtainable in packets of assorted colours from most stationers.

The weekly examination, with use of clips, ensures accuracy in the permanent records. For example, if in the first week one female of a trio is seen to be pregnant and her card is clipped to show this,

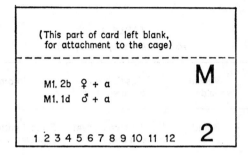

(a)

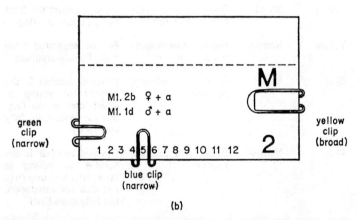

(b)

Figure 2. *Cage cards* (a) *Lay-out of card for mating M2.* (b) *Position of clips* shows second litter partly classified; male absent (e.g. to avoid a further litter) disease suspected (e.g. infantile diarrhoea).

then in the second week when a new litter is seen in the cage, it is clear that the litter belongs to the female with the clipped card.

Four coloured clips are amply sufficient to convey all the current information one needs. Each clip can be put on the card with either the broad or the narrow side visible, giving eight kinds of information altogether. Table 4 shows a well-tried way of using them.

TABLE 4

Information supplied by clips

Colour	Visible side	Position on card	Information
Red	Broad	Right, between symbol and number	Female with male. Litter expected within a week
Red	Narrow	Right, between symbol and number	Female with male. Female probably pregnant
Yellow	Broad	Right, between symbol and number	Female separated from male. Female nursing
Yellow	Narrow	Right, between symbol and number	Female separated from male. Female pregnant
Blue	Broad	Base, over relevant number	Covered number is the number of young in litter; litter is old (e.g. due for separation, fully classified)
Blue	Narrow	Base, over relevant number	Covered number is the number of young in litter; litter is young (e.g. not due for separation, not fully classified)
Green	Broad	Left, below reference nos., etc. of pair	Cage should be inspected daily (e.g. female is a poor mother, and litter may need fostering)
Green	Narrow	Left, below reference nos., etc. of pair	Disease suspected in litter or adults. Decision on culling required

Preserving techniques and chromosome preparations

Preserved specimens and progenies, and chromosome preparations, provide additional data or information.

(a) Preservation techniques

Adult specimens of certain phenotypes may be preserved by skinning (the curator of a zoology museum can often advize on procedure). These can be typical specimens of all of the phenotypes of all generations involved in a completed investigation whose records are to be used in the future for study; alternatively, a simple pedigree may be made in diagrammatic form, using skinned specimens, for the parental and F1 generations of a simple investigation, and used as an adjunct to data for the F2 provided either as records or as frozen progeny.

For large numbers of mice, e.g. a whole segregating progeny, it is more convenient to use young ones (because of their small size and ease of storage) and to preserve them by deep-freezing (Wallace, 1965). The whole progeny, slowly thawed, can then be used for a class exercise or for a problem in a practical examination. Deep-frozen progenies can, with care, be used ten times or more.

Where colour is unimportant, as for skeletal or hair-texture mutants, specimens can be preserved and arranged in a permanent life-like position by impregnation with certain chemicals (apply to Harris Biological Supplies Ltd., Oldmixon, Weston-super-Mare, Somerset, who also sell such specimens).

(b) Chromosome preparations

As an adjunct to work on independent segregation and linkage, it is useful to be able to see the twenty pairs of chromosomes from mice one has actually bred. Karyotypes from normal mice can best be made from preparations of corneal tissue (Fredga 1964).

References

Wallace, M. E. (1965). 'Using Mice for Teaching Genetics'. *Schools Science Review*, nos. 160 and 161.

Luker, A. J. and Luker, H. S. (1970). *Laboratory Exercises in Zoology*. Butterworth.

Fredga, K. (1964). 'A simple technique for demonstration of the chromosomes and mitotic stages in a mammal. Chromosomes from cornea'. *Hereditas*, 51, 268–73.

4

Investigations with one locus

The possibility of variation between individuals in higher organisms is very large. Taking only one locus (i.e. gene site), and ignoring the many thousands of other loci existing in the mouse, one can readily see that there are three possible genotypes with two alleles: AA, Aa, aa. An easy way to obtain this list is to align the possibilities on one chromosome against those on the other, in a two-dimensional diagram:

	A	*a*
A	*AA*	*Aa*
a	*Aa*	*aa*

(The shaded portion contains a repeat of a genotype in the unshaded portion.) With three alleles, there would be six genotypes. This may be checked from a similar diagram, aligning, say, the three agouti locus alleles A, a^t, and a on one chromosome against the same three on the other chromosome. With four alleles, there would be ten genotypes. And so on.

Two-dimensional diagrams are a convenient way for working out in diploid organisms not only possible genotypes, but possible types of mating, as will be shown below. This is simply because reproduction involves two sexes. The same results may be obtained with algebraic expressions. (The relevant one above is $(A + a)^2$.) These are of course unlimited dimensionally, but they can be tedious to expand, whereas a two-dimensional diagram can show a lot of detail without complexity. For example, the information in the square spaces of the diagram above are deducible from the margins; these spaces need not therefore be filled in with genotypes, but left blank for some other information. This other information, as will be shown with further examples, can be in the form of symbols, figures, words, or shading.

Suppose, for example, a list of genotypes of mating, still with only two alleles at one locus, is required. One animal can be of three genotypes; these may be aligned against the three possible genotypes of its mate:

	AA	Aa	aa
AA			
Aa	//////		
aa	//////	//////	

From the two margins, the top left space is seen to be *AA* × *AA*, the next on the right *AA* × *Aa*, and so on; and the shaded spaces are repeats. Instead of filling these spaces *AA* × *AA*, *AA* × *Aa*, etc., one may write in further information, namely the usual name for these different types of mating:

	AA	Aa	aa
AA	truo	b.c.	o.c.
Aa	b.c.	i.c.	b.c.
aa	o.c.	b.c.	true

Thus there are: one genotype of outcross (o.c.), one of intercross (i.c.), two of backcross (b.c.), and two of matings between like homozygotes (labelled 'true' because, producing only further like homozygotes, they are said to breed 'true').

Investigation 1: Two alleles with no dominance

If the heterozygote differs in phenotype from both homozygotes, there is said to be no dominance between the two alleles involved. But if the heterozygote has the same phenotype as one of the homozygotes, the allele occurring in that homozygote is said to be dominant to the other allele. Thus there is no dominance between the *A* and *a*[t] alleles, but full dominance between the *A* and *a* alleles. (See the description of the genotypes *AA*, *Aa*[t], *Aa* and *aa* in Table 1, Chapter 1.)

Alleles without dominance are useful for studying the gametic output of an individual, i.e. for listing the genotypes of an individual's gametes and their frequencies. This is because the phenotype of each genotype in the progeny is unique, and so one has only to recognize and count the actual phenotypes produced, in order to deduce the genotypes of the parents' gametes and their frequencies. In fact, the only way to observe exactly the flow of gametes from one generation to another in a diploid organism is to use alleles without dominance. Further, in so far as two such alleles label or 'mark' the two homologous chromosomes on which they are sited, a large part of what happens to these chromosomes at meiosis may also be deduced. (The rest requires cytological information.)

A simple way to start this study is to mate the heterozygote to one of the homozygotes, for then the number and frequency of phenotypes in the progeny refers to the segregation of gametes in only one parent, the heterozygote, the homozygote being able to produce only one kind of gamete. The last diagram, were the a^t allele to be substituted for the a allele, would show that, for the agouti alleles, there are two equally suitable matings: the backcrosses $Aa^t \times AA$ and $Aa^t \times a^t a^t$. If the first produces light-bellied and dark-bellied agoutis in equal numbers (or nearly equal, allowing for chance), then the heterozygote Aa^t must have produced gametes A and a^t in equal numbers; similar reasoning applies to the second.

Matings of heterozygotes to homozygotes are often termed 'testcrosses' (instead of 'backcrosses') because they 'test' the gametic genotype of the heterozygote. They can of course be made with a great number of mutants in mice. *All* such matings produce two phenotypes of offspring in equal numbers. (The only exception known at present is the '*t*-locus', where very unusual mechanisms are at work.)

A more complex way to start this study is to mate two heterozygotes together; it is also a more interesting one as it prepares for an understanding of segregation with dominance. The genotypes of offspring expected from such an intercross can be listed by aligning the gametes from one heterozygote, with their frequencies, against the gametes from the other:

	A	a^t
	1 :	1
A 1	1	1
a^t 1	1	1

1 AA: 2 Aa^t: 1$a^t a^t$, i.e. three phenotypes in the ratio 1 : 2 : 1.

This is one of the familiar 'monohybrid' ratios. It was so called because it was discovered by intercrossing the F1, or 'hybrids', from a cross between members of two true-breeding stocks differing at one locus.

The programme of matings carried out in typical pre-Mendel hybridizing experiments is suitable, in fact, for exploring completely all the possibilities involving two alleles without dominance. The terms 'outcross', 'backcross', and 'intercross' were used originally in hybridizing experiments and referred to the relationships between mates. Thus the first referred to a cross to something true-breeding 'outside' some original true-breeding stock; the second to a cross between their progeny, the hybrid, 'back' to one of the two true-breeding stocks; and the third to the crossing *'inter' se* of the hybrids. These terms then came to mean the genotypes of mating underlying these relationships. And they are now more often used to mean genotypes of mating irrespective of relationship.

Such a full programme is shown below, together with lines showing the expected passage of each allele from generation to generation. The alleles chosen are c^e (extreme chinchilla) and c (albino) (Plate 5), although A and a^t are just as suitable. The former have the practical advantage that the three phenotypes

Generation	Genotypes (mates are shown in pairs on the same line)	Name of mating (by relationship and genotype)
	c^ec^e c^ec^e cc cc	true
Parental (P)	c^ec^e_____cc	outcross
First filial (F1)	c^ec_____c^ec	intercross
Second filial (F2)	c^e : 1 c^ec^e : 1 c^e c 1 c^ec c^ec 1 c cc	

involved can be classified by the colour of the eyes and coat without much handling, whereas belly-colour in agouti-locus phenotypes can only be seen by handling. The two backcrosses $c^e c^e \times c^e c$ and $cc \times c^e c$ are not shown; classically they are the mating of F1 to each P generation, but matings of these genotypes can clearly be made between members of other generations.

The diamond-shaped diagram above may now be tipped from its point on to one side. The margins imply the genotypes expected in the spaces, and their phenotypes can be written there instead: The marginal frequencies are then multiplied to give:

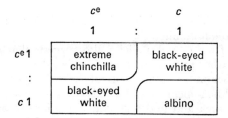

1 extreme chinchilla : (1 + 1) black-eyed white : 1 albino = 1 : 2 : 1, or algebraically $(1c^e + 1c)^2$.

Investigation 2: Two alleles with dominance

It may be noticed that in the last investigation, the heterozygote is intermediate only inasmuch as it resembles one homozygote in one part of its body and the other in another. This is a little unusual, for in most organisms with intermediate heterozygotes, only one part is concerned (e.g. the heterozygote for long and short ray in *Senecio vulgaris* has a half-length ray).

However, this feature of the albino locus is useful for it shows what is to be expected for alleles with full dominance. If eye-colour were ignored in the last diagram, the $c^e c$ and cc genotypes would be similar, i.e. white-coated, and this would be the same as saying that c is dominant to c^e; the diagram would then become as shown below (left), giving a ratio 1 fawn : (2 + 1) white. And if coat-colour were ignored in the last diagram, the $c^e c^e$ and $c^e c$ genotypes would be similar, i.e. dark-eyed, and this would be the same as saying that c^e is dominant to c; the diagram would then become as shown below (right), giving a ratio $(1 + 2)$ black-eyed : 1 pink-eyed.

This is, then, the other familiar 'monohybrid' ratio, 3 : 1. The

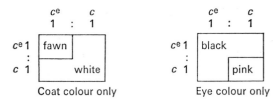

Coat colour only   Eye colour only

above exercise shows how it is derived from the 1 : 2 : 1 pertinent to alleles without dominance.

The use of the c^e and c alleles also suggests that dominance depends on what part of the body is affected by a gene. There are many other mutants whose normal alleles show full dominance, but it should be remembered that this dominance does refer only to a certain area and under certain limitations of observation. Thus although black, $+$ or B, is said to be fully dominant to brown, b, this statement applies only to the coat pigment; in poor light the eyes of bb look the same colour as those of BB, but in a good light they appear dark ruby, not black; and unpigmented areas, like the intestines, show no difference between BB and bb.

The brown mutant (Frontispiece) is very popular for investigations into full dominance. For comparison with the diagrams for c^e and c, the diagram showing how to work out the progeny to be expected from $Bb \times Bb$ is given below:

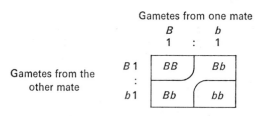

Thus the expected frequency is 3 black : 1 brown. A useful investigation is, then, to do the whole classical programme as

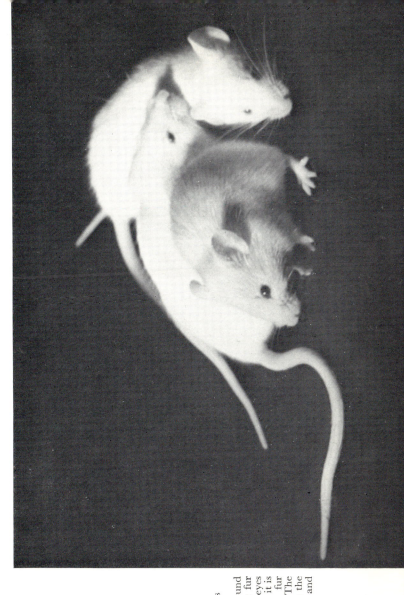

Plate 5 Some albino locus phenotypes

The mouse in the foreground is extreme chinchilla, $c^e c^e$: fur coloured (pale fawn), eyes black. The mouse behind it is black-eyed white, $c^e c$: fur nearly white, eyes black. The mouse at the back and to the right is albino, cc: fur white and eyes pale pink.

Plate 6 The short-ear and normal phenotypes
The mouse on the left is short-ear, *sese*; compare with its long-eared companion. (The mouse on the right is normal at all loci; compare its plain sleek coat with those of tabby mice, Plate 6, and wavy-haired mice, Plate 7.) [Reproduced by permission of the *Journal of Biological Education*.]

above: cross true-breeding blacks to true-breeding browns, inter-cross the black F1 to obtain blacks and browns in F2 in the 3 : 1 ratio, and backcross black F1 to browns from the true-breeding stock (or those obtained in F2) to form the 'backcross' or 'testcross' mating.

An interesting addition to the programme is to test members of the F2 blacks by backcrossing them to browns to see whether, as is expected, some of them are heterozygous and others homo-zygous.

A complete list of available outcross mates, suitable for this type of full programme, is given below.

$$\left. \begin{array}{c} AA \\ aa \end{array} \right\} \qquad \left. \begin{array}{c} ++(AA) \\ p\ p\,(AA) \end{array} \right\} \qquad \left. \begin{array}{c} ++(aa) \\ c\ \ c\,(aa) \end{array} \right\}$$

$$\left. \begin{array}{c} a^t a^t \\ aa \end{array} \right\} \qquad \left. \begin{array}{c} ++(AA) \\ se\ se\,(AA) \end{array} \right\} \qquad \left. \begin{array}{c} ++(aa) \\ c^e\ c^e\,(aa) \end{array} \right\}$$

$$\left. \begin{array}{c} ++(aa) \\ p\ p\,(aa) \end{array} \right\} \qquad \left. \begin{array}{c} ++(aa) \\ b\ \ b\,(aa) \end{array} \right\}$$

$$\left. \begin{array}{c} ++(aa) \\ se\ se\,(aa) \end{array} \right\} \qquad \left. \begin{array}{c} ++(aa) \\ d\ \ d\,(aa) \end{array} \right\}$$

Investigation 3: Two alleles with lethality

Most mouse mutants are recessive. Less than ten of those mutants usually given a capital letter are fully dominant. Most of them, strictly, lack dominance, for the heterozygote differs from both homozygotes, because one is lethal. Thus, Danforth's short-tail, $Sd+$ (Plate 7), differs from normal ($++$, long-tail) and from $SdSd$ because the latter has no tail and dies before birth (or within a few days of birth). However, since the heterozygote can be crossed to only one of the homozygotes, the normal one, and then produces some mutant progeny, the term 'dominant' has often been applied to the mutant. Since symbolism for alleles lacking dominance is sometimes confusing, it has become convenient to give mutants whose homozygote is lethal a capital letter.

A simple investigation into this situation is to cross two short-

D

tails together: their genotype can only be $Sd+ \times Sd+$ and the progeny expected are:

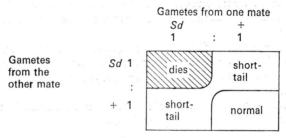

i.e. two short-tails to one normal.

Danforth's short-tail has the advantage that the mutant homozygote occasionally survives long enough to be seen, thus supporting the supposition that the observed 2 : 1 ratio is merely $(3-1) : 1$. In the case of the yellow mutant, A^y (Frontispiece), the homozygote never survives to be seen. In this case, a convincing way to do the investigation is to mate yellows together for several generations in an attempt to 'breed true', i.e. to produce only yellows. This will never be possible, because an all-yellow progeny requires that at least one mate be $A^y A^y$. Thus in every generation, non-yellows will appear; it should be checked over all generations that the ratio is 2 yellow : 1 non-yellow. Suitable yellows are: either $A^y a^t \times A^y a^t$ or $A^y A \times A^y A$; in the former investigation the non-yellows will be black-and-tans, and in the latter agoutis.

5

Investigations with two independent loci

If there are alternative alleles at two loci instead of at one, the possibility of variation between individuals is greatly increased. With hypothetical alleles g and *g* at one locus, and h and *h* at another, all possible genotypes may be obtained by aligning the possibilities at one locus against those at the other, i.e. gg, g*g*, and *gg* against hh, h*h*, and *hh*, in a 3 × 3 diagram (or by the appropriate algebraic expression). That is, there are 9 genotypes.

A convenient way to produce them all, and to find what segregations the more interesting matings can give, is to repeat the programme described for one locus, but choosing outcross mates differing at *two* loci on different chromosomes. That is, in terms of the g and h loci, outcross gg*hh* × *gg*hh, intercross the doubly heterozygous F1, g*gh*h, to raise an F2, and backcross the doubly heterozygous F1 to the double recessive, *gghh*, which will appear among the F2. Because the true-breeding stocks from which the outcross is made differ for two characters, this has been called a 'dihybrid' programme and it will give the familiar 'dihybrid' ratios.

It is logical to consider first two loci each having alleles which lack dominance. Although the *c* and *A* loci alleles qualify, they are not suitable for this programme, for reasons which will become clear. It will be assumed, therefore, that g and *g*, and h and *h* lack dominance, and the expected outcome of matings involving them will be given in theory. The student can see for himself retrospectively how, given phenotypes reflecting unique genotypes – as with alleles lacking dominance at *one* locus – one may deduce the genotypes, with their frequencies, of the gametes which produced them. This exercise, as with that for one locus, is worth pursuing, as it prepares for an understanding of dihybrid experiments with full dominance.

The first interesting mating is the double intercross: the *inter se* mating of F1, gg*h*h × gg*h*h. Each F1 produces gametes g and *g* in equal numbers, and gametes h and *h* in equal numbers; as non-homologous chromosomes are involved, these can combine at random, so that four genotypes of gamete are produced in equal numbers. The algebraic expression is $(1g + 1g)(1h + 1h) =$ 1gh, 1g*h*, 1*g*h, and 1*gh*; or again, the appropriate two-dimensional diagram would give the same result.

Conventionally, and perfectly logically, the next step is to obtain all possible genotypes in F2, and their frequencies, by aligning the four genotypes of gamete from one parent against those from the other, with their frequencies, and thus to obtain the following table of genotypes:

		Gametes from F1♂			
		gh 1	*gh* 1	g*h* 1	*gh* 1
		:	:	:	:
	gh 1 :	1 gghh	1 ggh*h*	1 g*g*hh	1 g*g*h*h*
Gametes from F1♀	*g*h 1 :	1 g*g*hh	1 g*g*h*h*	1 g*g*hh	1 g*g*h*h*
	gh 1 :	1 g*g*hh	1 g*g*hh	1 g*g*hh	1 g*g*hh
	gh 1	1 g*g*hh	1 g*g*h*h*	1 g*g*hh	1 g*g*hh

From this table, identical genotypes are marked and counted in some way, and the resultant ratios arranged in some meaningful manner, e.g.:

gghh	ggh*h*	gghh	ggh*h*	g*g*h*h*	*gg*hh	gg*h*h	gghh	*gghh*
1 :	1 :	2 :	2 :	4 :	2 :	2 :	1 :	1

However, this way of obtaining the results is laborious and not particularly useful. A simpler way, and therefore one less open to mistakes in the making, is to use the 3 × 3 diagram mentioned in the first paragraph of this chapter. Its usefulness will become apparent throughout the rest of the chapter.

The genotypes at the g and h loci are aligned against each

other, each with the frequencies appropriate to segregation in F2 at one locus: gg, g*g*, and *gg* will be obtained in a ratio 1 : 2 : 1, and so will hh, h*h*, and *hh*, for this is the monohybrid ratio. The full genotypes, i.e. for these loci taken together, may then be read from the margins. The frequencies of these full genotypes may be obtained by multiplying the frequencies in the margins (since it is assumed that the loci are on different chromosomes and so segregate independently). In other words, the diagram consists of the expected F2 segregation at the g locus and at the h locus *separately*, written in the margins, and the body of the diagram is the expected segregation of F2 for both loci taken *together*. Since the full genotypes can now be read easily from the margins and each genotype now occurs only once (in contrast to the former diagram), it is convenient not to write them in the spaces, but simply to write in the frequencies.

		F2 genotypes at g locus		
		gg 1 :	g*g* 2 :	*gg* 1
	hh 1	1	2	1
F2 genotypes at h locus	h*h* 2	2	4	2
	hh 1	1	2	1

It can readily be checked that the content of this diagram gives the same list of nine genotypes, and in the same frequencies, as appears above. The diagram can now be used, with appropriate substitutions for the alleles, to obtain expectations in the F2 for a number of dihybrid investigations.

A diagram for a backcross of the F1 (g*g*h*h*) to the double recessive (*gghh*) is as simply obtained: align the F2 genotypes at the g locus (g*g* and *gg* in frequencies 1 : 1) against those for the h locus (h*h* and *hh* in frequencies 1 : 1), giving 1 g*g*h*h* : 1 g*ghh* : 1 *gg*h*h* : 1 *gghh*.

Investigation 4: Two pairs of alleles each with full dominance.

Two mutants very suitable for this investigation are short-ear, *se* (Plate 6), and pink-eyed dilution, *p* (Frontispiece), for both are fully recessive to their normal alleles. From true-breeding stocks of each, make the outcross $pp++$ × $++sese$ to obtain the doubly heterozygous F1, $+p+se$; intercross these to obtain the F2 segregation; and make a backcross of F1 $+p+se$ mated to $ppsese$ when these are produced in the F2.

The frequencies of genotypes expected in F2 are obtained by substituting p for g and se for h in the 3 × 3 diagram above. However, since full dominance is now involved, this diagram does not represent the phenotypes. But these are simply obtained by erasing for each locus those lines within the body of the diagram which divide the homozygous normal from the heterozygote. This groups together genotypes of similar phenotype:

F2 genotypes at p locus

		++	+p	pp
		1 :	2 :	1
F2 genotypes at *se* locus	++ 1	1	2	1
	+se 2	2	4	2
	sese 1	1	2	1

Adding the frequencies of similar genotypes, there are now: 9 fully normal : 3 short-eared : 3 pink-eyed dilute : 1 short-eared pink-eyed dilute. This familiar dihybrid ratio 9 : 3 : 3 : 1 is simply $(3+ : 1p)$ $(3+ : 1se)$, where single symbols $(+, p, se)$ stand for phenotypes rather than genotypes. In fact, if a list of genotypes is not required, but simply one of phenotypes, this may be obtained by writing phenotypes and their frequencies in the margins:

F2 phenotypes at p locus

		+	p
		3 :	1
F2 phenotypes at *se* locus	+ 3	9	3
	se 1	3	1

In a similar manner it can be shown that the genotypes and phenotypes from the double backcross are the result of a $(1+1)^2$ formula in place of this $(3+1)^2$ formula.

The use of p and se has the advantages that these loci affect different parts of the body and are classifiable at different ages. These allow one easily to ignore one locus and concentrate on the other. Thus, although for all generations both loci should be fully classified, one may collect and analyse the data for the p locus regardless of the phenotype at the se locus, and vice versa. In this way one can understand first the single-locus phenomena before having to tackle the combined segregations. As p is the earliest to be classified, this is clearly the locus to concentrate upon to begin with. These properties of the p and se loci also expand one's ideas of what 'independence' means: for they are physiologically independent as well as segregating independently.

Several of the genotypes listed in the last chapter, in Investigation 2, are useful for dihybrid studies as well as monohybrid. The complete list of suitable outcross mates available is:

$$\left.\begin{array}{l} a\ a + + \\ AA\ se\ se \end{array}\right\} \quad \left.\begin{array}{l} + +\ se\ se \\ p\ p + + \end{array}\right\} \begin{array}{l} \text{both homozygous} \\ \text{for } A \text{ or for } a \end{array} \left.\begin{array}{l} AA + + \\ a\ a\ p\ p \end{array}\right\} *$$

$$\left.\begin{array}{l} d\ d\ a\ a \\ + + AA \end{array}\right\}* \quad \left.\begin{array}{l} AA + + \\ a\ a\ b\ b \end{array}\right\}* \quad \left.\begin{array}{l} a^{t}a^{t} + + \\ a\ a\ p\ p \end{array}\right\}*$$

$$\left.\begin{array}{l} d\ d\ a\ a \\ + + a^{t}a^{t} \end{array}\right\}* \quad \left.\begin{array}{l} a^{t}a^{t} + + \\ a\ a\ b\ b \end{array}\right\}* \quad \left.\begin{array}{l} AA + + \\ a\ a\ se\ se \end{array}\right\}*$$

$$\left.\begin{array}{l} a^{t}a^{t} + + \\ a\ a\ se\ se \end{array}\right\}*$$

Those marked with an asterisk * do not conform to the conventional way of doing dihybrid investigations, for the mates do not differ each for a different mutant; rather, one mate is recessive for both loci and so differs from its mate for two mutants at once. However, this makes no difference to the segregations in each generation; and in programmes using these outcross mates, the double backcross may be made earlier.

Investigation 5: Two pairs of alleles with dominance and epistacy

It may be wondered why the albino gene has been excluded from the list of mates suitable for the investigations in the last section.

True, it is fully recessive, and there are loci on other chromosomes with which it could be made to segregate. However, it has been excluded because phenotypically albino mice can have many genotypes, and few are suitable as mates for the genotypes listed in that section. Here are some of the possible genotypes of albinos, taking for convenience the first few letters of the alphabet which are also gene symbols: *aabbccdd*, *aa++ccdd*, *aabbcc++*, *AAbbccdd*, *Aa+bcc+d*, *++++cc++*. All these genotypes have the one phenotype, albino. However, if the colour gene (+, in place of *c*), were present, the genotypes would all have differently coloured phenotypes (except the last two which would both be normal). In other words, in *cc* mice, none of the other colour loci can express themselves in the phenotype: albinism is said to 'stand upon', i.e. to be *epistatic* to, other such loci. The capacity of albino mice, unless of known and indeed specified ancestry, to 'hide' other colours, explains their sometimes startling tendency to produce unexpected phenotypes in later generations, thereby making many an ill-advized experiment 'go wrong'.

The classic way to study epistacy is to cross an individual homozygous for one recessive colour mutant to an albino known to be genetically different from the first individual only in respect of albinism and the other colour locus; the other locus should also be on a different chromosome from that of albinism. Then proceed as in the previous programmes in this chapter: intercross from the outcross, and backcross the F1 to the double recessive. The usual dihybrid ratios will be modified; and from the way in which they are modified, the nature of epistacy becomes apparent.

Taking as an example the outcross *cc++* × Maltese dilution *++dd* (both homozygous *aa*) (Plate 5 and Frontispiece), the first segregating generation, F2, is modified as follows. Substituting *c* for *g* and *d* for *h* in the 3 × 3 diagram for g and h, one has the genotypes and their frequencies for the *c* and *d* loci. To obtain the phenotypes, erase the line dividing +*c* from its homozygote ++; the line dividing +*d* from its homozygote ++; and finally the line dividing +*d* from *dd* in the column headed *cc*; as in the diagram below. There are three phenotypes instead of four, namely 9 coloured : 3 diluted : 4 albino. It may now be checked that the albinos are of three genotypes at the *d* locus, by mating them to ++*dd* of the true-breeding stock from which the ++*dd* outcross

F2 genotypes at *c* locus

		++	+c	cc
		1 :	2 :	1
	++1	1	2	1
F2 genotypes	:			
at *d* locus	+d 2	2	4	2
	:			
	dd 1	1	2	1

mate was obtained. Indeed this will be necessary in order to identify which albino is the double recessive *ccdd* needed for the backcross mating. The outcome of all possible genotypes of mating albino × ++*dd* can easily be worked out, the one involving *ccdd* being the only one to produce nothing but dilutes.

The genotypes produced by the backcross mating +*c*+*d* × *ccdd* are as simply obtained, namely equal numbers of +*c*+*d*, *cc*+*d*, +*cdd*, and *ccdd*. The phenotypes, however, will not be in equal numbers, there being one normal : one dilute : two albinos. The epistacy of albinism can again be demonstrated by crossing these albinos to ++*dd* and obtaining progeny indicating two kinds of albino.

The following is a complete list of available outcross mates suitable for the programme above:

$$cc++(aa) \atop ++dd(aa) \Big\} \qquad\qquad c\ c\ a\ a \atop ++AA \Big\}*$$

$$cc++(aa) \atop ++bb(aa) \Big\} \qquad\qquad c\ c\ a\ a \atop ++a^t a^t \Big\}*$$

Investigation 6: Mimic genes and interaction

Mutant genes at different loci which give the same phenotype are called mimics. Thus waved-2 and wellhaarig (*wa-2* and *we*) produce a curly-haired curly-whiskered mouse, whether the waved-2 mutant is homozygous or the wellhaarig one is homo-

* See the note at the end of Investigation 4.

zygous (Plate 7). For some mimic genes the double mutant is phenotypically different from the two single mutants, for others they are the same; in either case the ratios obtained in the various generations from a programme such as the one used throughout this section, are modified from the classical ones.

To take the waving genes as an example: the outcross would, like all outcrosses so far considered, produce normal F1 progeny; but this at first sight might seem odd, for in the ordinary way two wavies mated together would be expected to breed true. In this case, the outcrossed wavies would be $++wa\text{-}2wa\text{-}2$ and $wewe++$, and their offspring $+we+wa\text{-}2$, i.e. normal. The first demonstration of mimicry is then, the breeding of two stocks of wavies, each separately breeding true, and the crossing of one to the other, when they cease to breed true.

The F1 $+we+wa\text{-}2$, crossed *inter se*, would of course give the 9 genotypes in ratios: $1 : 1 : 2 : 2 : 4 : 2 : 2 : 1 : 1$, familiar in the dihybrid situation; but the phenotypes would not occur in the familiar $9 : 3 : 3 : 1$ ratio – in fact there would be less than four phenotypes, for in this case all individuals homozygous for one or more wavy genes would look the same. This may be seen by substituting $wa\text{-}2$ for g and we for h in the 3×3 table for g and h; then erase all the lines except those dividing $+wa\text{-}2$ from $wa\text{-}2wa\text{-}2$ (as far as the $wewe$ row) and those dividing $+we$ from $wewe$ (as far as the $wa\text{-}2wa\text{-}2$ column):

		F2 genotypes for wa-2 locus		
		++	+wa-2	wa-2wa-2
		1	: 2 :	1
F2 genotypes for we locus	++ 1 :	1	2	1
	+we 2 :	2	4	2
	+wewe 1	1	2	1

The expected F2 ratio is thus 9 normal : 7 wavy.

Identification of the double recessive, needed for the backcross, would require quite a complicated programme, possibly not feasible in practice. It can best be tackled first therefore as a paper problem, together with demonstration of the expected outcome of the backcross.

There are many types of interaction between genes at different loci. When the double mutant looks much as one would expect by adding the effects of each single mutant, interaction is said to be additive. This is the case for all the pairs of mutants listed as suitable for Investigation 4 (p. 43). Two examples of non-additive interaction, epistacy and mimicry, have been given in Investigations 5 and 6; there are many others, and nearly all of them result in phenotypic ratios in F2 different from the classical 9 : 3 : 3 : 1. Prediction both of phenotypes and of frequencies in F2 is not therefore a simple matter of seeing the effect of each mutant separately and then extrapolating.

6

Simple investigations with linkage and limitation

Some genetic phenomena are popularly confused. A few will be described in this chapter; their distinguishing features will become clear.

Investigation 7: Sex linkage

In all but a few mammalian species, an individual carrying a particular pair of large chromosomes called 'X' chromosomes, is a female, and an individual carrying one of these and a much smaller one, called the 'Y', is a male. These 'sex chromosomes', however, are concerned also with heritable characters quite unconnected with sex. These are borne almost without exception on the X chromosomes: the Y is so small that it has no counterpart to the gene-bearing portion of the X.

Two features follow from the last statement. All sex-linked genes carried on the X cannot be carried on the Y; and an X-linked mutant has the same phenotype when homozygous in the female as when it is carried by the single X of a male. It follows also that a sex-linked mutant can be heterozygous only in the female. When 'dominant', it has a different phenotype in the heterozygote from that in the homozygote, for the former carries a normal gene which is expressed in the phenotype along with the expressed mutant effect. Therefore these 'dominant' X-linked mutants are, strictly speaking, misnamed; however, it has become convenient to give those which produce mutant phenotypes on outcrossing a capital letter.

Dominant X-linked genes in the mouse are not on the whole easy to classify. The best from this viewpoint is tabby, *Ta*, but it is wise to avoid crossing outside the stock available commercially. Apart from normal females, the possible genotypes, and their phenotypes, are (see Table 1, Chapter 1, and Plate 8):

	X^{Ta}	X^+
X^{Ta}	greasy ♀	striped ♀
Y	greasy ♂	normal ♂

A conventional way of writing the female genotypes is: TaX/TaX and $TaX/+X$, and the male: $TaX/+Y$ and $+X/+Y$. However, since Ta cannot get on to the Y, it is legitimate to let the Ta symbol stand for a Ta-bearing X, and the symbol X for a normal-bearing X: then these four genotypes become simply $TaTa$, TaX, TaY, and XY.

In man the best-known sex-linked mutant is haemophilia – best-known because it affected a number of the princes of Europe; when it was investigated, all such men were traced to Queen Victoria, in whom the mutation must have occurred. The full pedigree (Stern, 1960) shows what was then popularly called 'criss-cross' inheritance, namely that a haemophiliac man derives the mutant gene from his mother and passes it to his daughters.

This 'criss-cross' situation is best demonstrated with the Ta gene by crossing, in alternate generations, a striped female to a normal male and a normal female to a greasy male:

Generation		Passage of Ta
1	$TaX \times XY$	♀
2	$TaX \quad XX \times TaY \quad XY$	♂
3	$TaX \times XY$	♀

It is then clear that a greasy male derives the mutant from his mother and passes it to his daughters. Put another way, and given a long enough pedigree, one could say: find the most abnormal phenotype (greasy), and an abnormal one (striped) will be found, up and down a generation, in the *opposite* sex.

For practical purposes, note that the Generation 1 mates are available in Generation 2 (shown by the dotted lines); if the stock becomes difficult to continue exactly as above, because in any one generation the greasy males breed poorly, intermediary repeat matings $TaX \times XY$ may be made. Similarly a check on the equivalence of the $TaTa$ and TaY phenotypes can be made by mating TaX $\times$ TaY in Generation 2 in order to obtain $TaTa$ (one-quarter of the progeny).

Investigation 8: Sex limitation

Sex limitation is a phenomenon readily confused with sex linkage. Yet its most prominent feature is that only one sex is affected (or, sometimes, one sex is regularly more affected than the other). If it is the male sex which is affected generation after generation, the phenomenon cannot be sex linkage for, as can be seen above, an affected male cannot pass his X-linked defect to his sons. If the female sex is affected, or affected more strongly, generation after generation, then, especially if mating is at random (i.e. there is no close inbreeding), it cannot be X-linkage; for an X-linked mutant in single dose (there would be very few homozygotes without inbreeding) gives a more abnormal phenotype in the male than in the female.

A clear definition of sex limitation will show how it accounts for a predisposition of one sex to be affected rather than another. It is the phenomenon whereby the expression of a mutant carried on an autosome (i.e. any chromosome but an X or Y) is conditioned by the sex of the individual. Sexual differentiation affects a number of organs and tissues in mammals; it should not therefore be surprising that autosomal mutants also affecting them should be limited in expression by their developed nature. Sex, seen as a heritable character, simply interacts with the autosomal mutant. To consider an extreme case, a mutant affecting milk production must be limited in expression to females, even when it occurs in males.

In mice the urogenital systems of the two sexes are arranged somewhat differently, and it is here that sexual limitation of an autosomal gene is fairly easily observed. In a stock breeding true for short-ear, *se* (Plate 6), about half the male mice on autopsy would be found to have hydronephrosis ('watery kidneys'), and only about one in ten of the females (Wallace and Spickett, 1967). It is not known exactly how this gene affects the development of the kidneys, but in this situation the mutant responsible is clearly occurring equally frequently in both sexes, yet one sex is far more affected than the other.

Since the 'wateriness' of the kidneys advances with age, a practical investigation would plan autopsy of mice at least ten months old. Since hydronephrotic males are often too severely affected to breed well, it might also be more practical not to plan long-term *sese* × *sese* matings. Rather, simply allow all those matings involving short-ear in other programmes (see Chapters 4, 5, and 7) to run to infertility, and keep for at least ten months all *sese* progeny not wanted for breeding but for which space can be found. Then autopsy and dissect all *sese* mice.

Investigation 9: Close autosomal linkage

Close linkage is shown with *Ta*, for this mutant never leaves the X. Autosomal linkage is commoner than sex-linkage simply because there are more autosomes than sex chromosomes. Linkage may be loosely defined as the tendency of two mutants at different loci to remain together (or to remain apart) generation after generation, if they came initially from the same parent (or different parents). Historically the observation of this phenomenon led to the terms 'coupling' for the 'together' situation and 'repulsion' for the 'apart' (indicated in brackets above). It was not known that they were facets of the same phenomenon until the relation between mutants and chromosomes was discovered; then it was realized that only mutants with loci in the *same* homologous pair could exhibit the phenomenon. A double heterozygote is now said to show the coupling phase of linkage if the mutants are on one homologue and their normal alleles on the other, and to show the repulsion phase if the two mutants are on different homologues – and their normal alleles thus also on different ones. This may be represented

$$\frac{+\ +}{c\ \ p} \quad \text{and} \quad \frac{+\ \ p}{c\ \ +}$$

respectively, where c and p stand for any linked mutants.

The actual mutants albino, c, and pink-eyed dilution, p, are, in fact, linked. Their linkage, plus the epistacy of albinism, is further reason why many an experimenter intending to investigate dihybrid inheritance has, ill-advizedly using these mutants, obtained quite inexplicable results. Epistacy precludes the use of albinism in linkage experiments involving colour loci.

Two mutants used in previous chapters are, in fact very suitable for linkage studies. Maltese dilution, *d*, and short-ear, *se* (Frontispiece and Plate 6) are very closely linked, so closely in fact that they almost never change from their initial relationship. Thus if short-ear intense mice are mated to long-ear dilute mice, generations of further breeding without crossing outside this stock will not produce a short-ear dilute mouse; while if mice already short-ear dilute are crossed to normal ones, and further generations bred without crossing outside this stock, every short-ear mouse will be dilute and every long-ear mouse intense. When they do break away from their relationship, it is because 'crossing-over' has occurred, i.e. the homologues have broken between the *se* and *d* loci and joined up, one end of one with the other end of the other. Thus:

(This is diagrammatic: the full cytological situation is more complex.) The reason why crossing-over almost never occurs is that the *se* and *d* loci are so close that there is barely room for this to happen.

This being the case, an investigation of close linkage requires, in order to demonstrate both phases, two initial crosses. These are set out below, one above the other, but they can be carried out simultaneously. Note that double heterozygotes are always crossed to the double recessive and that, for the repulsion programme, this is 'crossing outside the stock' (see above). The reason for using the double recessive will become clear in the next section. Note also that coupling backcrosses can be made by mating within the progeny in *each* generation after the first coupling one. On the other hand to continue producing repulsion backcrosses, one must alternate with each backcross a mating between single heterozygotes.

The phenomenon usually confused with close linkage is

Plate 7 Some mimic genes
Upper left: *wa-2wa-2 vtvt.*
Lower right: *wewe Sd+*. Note
that the waved-2 and wellhaarig genes each produces a
waved coat and curled whiskers,
and that vestigial-tail and
Danforth's short-tail each reduces tail length. These mice
are about twenty-eight days
old.

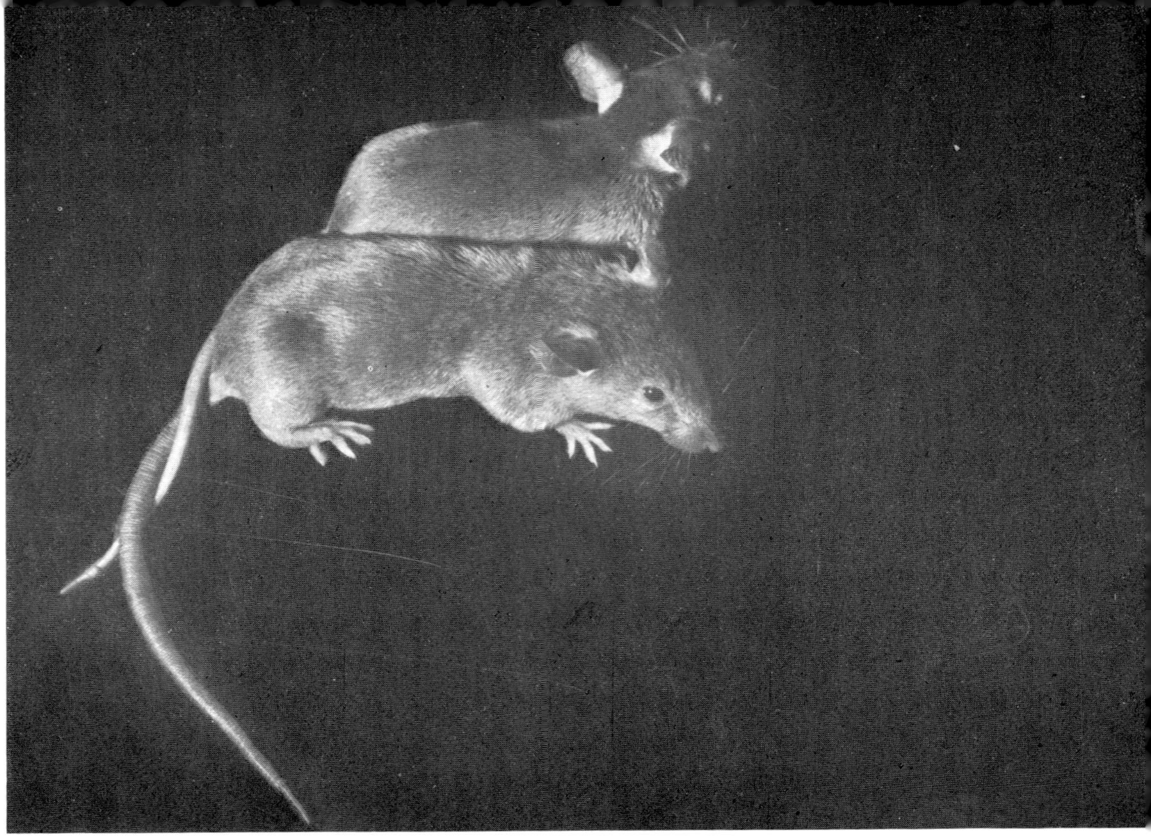

Plate 8 Two tabby phenotypes The upper mouse is a tabby male, *Ta*Y. Note the bare patch behind its ears, the dark dorsal area in its agouti fur, and the bare tail. The fur looks greasy and the tail feels greasy; the knob near the tip of the tail is not present in all specimens.

The lower mouse is a tabby female, *Ta*X. Note the stripes across her shoulders; facial stripes and the pronounced tail rings are not shown by all specimens.

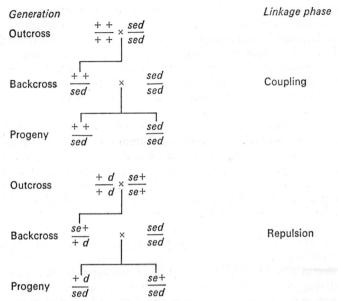

mutation. Confusion arises from the rarity of some crossovers; thus if the coupling backcross above were to produce a short-ear intense mouse, it might be thought that the *d* gene in the heterozygote had mutated to its normal allele. Strictly speaking, one such event could be interpreted either way, but if another short-ear intense appeared, crossing-over would seem more likely. For loci so close that crossing-over is as rare as mutation, doubt persists.

For each locus, however, less than 1 in 50 000 mice carries a newly arisen mutation. In most classroom situations where a mutation is suggested as an explanation of an unexpected phenotype, it is usually found that a mistaken mating has been made or that the mutants chosen for the work were unsuitable because of epistacy or some other phenomenon not known to the inexperienced.

Investigation 10: Loose autosomal linkage

When two loci are far apart on the same chromosome, there is clearly more chance for crossing-over to occur between them than when they are close. For this reason, the frequency of crossing-over is used as a measure of the distance between the loci. Thus if the phenotypes produced by crossing-over are common, the loci concerned are far apart; the extreme is when they are as common as those produced when no crossing-over occurs – then there would be doubt about the linkage, if there were no other information on this point.

Strictly speaking such products of crossing-over should not be called 'cross-overs'. The more usual term now is 'recombinant'. Thus, if short-ear dilute and fully normal mice were to appear in the progeny of a repulsion backcross (see last section), they would be called recombinants; and this is because they have resulted from an event which rearranges the chromosomal relationship in the double heterozygote. When crossing-over has not occurred, the chromosomal arrangement in the double heterozygote is maintained and the progeny phenotypes resulting are non-recombinants.

Again strictly speaking the gametes produced by the double heterozygote should be given these terms, not the progeny. Progeny phenotype results from the gametes from *both* parents, and it is wiser, given a choice, to study the direct products of a phenomenon whose frequency one wishes to measure. However, if a double recessive is used in matings made to measure linkage, the gametes of this double recessive will be the same genotype whether crossing-over occurs or not, so that progeny phenotype will reflect exactly the genotypes of gametes produced by the other (doubly heterozygous) parent. This is why backcrosses or 'testcrosses' are always preferred to, say, double intercrosses.

The proportion of a backcross progeny which is recombinant is termed the recombination value. For different pairs of loci in the mouse, this value ranges from 0% to 50%. Loose linkages are thus much commoner than close ones.

The mutants waved-2, *wa-2*, and vestigial tail, *vt* (Plate 7), are useful in that they show a recombination value half-way between the extremes, namely 25%. A good programme is to cross *wawavtvt* $\times$ $++++$ (*wa* will be written in place of *wa-2* from now on), and backcross the F1 double heterozygote to the double recessive. Classification of the progeny, which should preferably

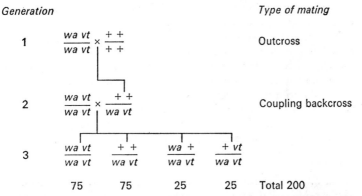

Generation

1    $\dfrac{wa\ vt}{wa\ vt} \times \dfrac{+\ +}{+\ +}$ Outcross

2    $\dfrac{wa\ vt}{wa\ vt} \times \dfrac{+\ +}{wa\ vt}$ Coupling backcross

3    $\dfrac{wa\ vt}{wa\ vt}$ $\dfrac{+\ +}{wa\ vt}$ $\dfrac{wa\ +}{wa\ vt}$ $\dfrac{+\ vt}{wa\ vt}$

    75   75   25   25   Total 200

be done soon after birth in order not to omit young which die early, and counting of similar phenotypes should then give figures which agree reasonably with the exact expectations (out of 200 progeny) given in the table above.

If the repulsion backcross is desired, there is no need to bring in a new stock as in the *se/d* programme, for the recombinants in Generation 3 above (the two on the right), if mated together (wavy to vestigial), will give a double heterozygote in repulsion and the double recessive necessary for the repulsion backcross. Note that in this type of mating crossing-over produces no new types of gamete: the progeny frequencies are therefore 1 : 1 : 1 : 1 and half of them are useful for the repulsion backcross. Classification and counting of backcross progeny should again give

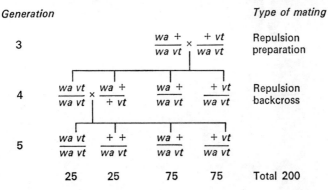

Generation

3    $\dfrac{wa\ +}{wa\ vt} \times \dfrac{+\ vt}{wa\ vt}$ Repulsion preparation

4    $\dfrac{wa\ vt}{wa\ vt} \times \dfrac{wa\ +}{+\ vt}$ $\dfrac{wa\ +}{wa\ vt}$ $\dfrac{+\ vt}{wa\ vt}$ Repulsion backcross

5    $\dfrac{wa\ vt}{wa\ vt}$ $\dfrac{+\ +}{wa\ vt}$ $\dfrac{wa\ +}{wa\ vt}$ $\dfrac{+\ vt}{wa\ vt}$

    25   25   75   75   Total 200

Type of mating

figures in reasonable agreement with the exact expectations (out of 200 progeny) given on page 55.

Note that the ratios of the phenotypes in the repulsion backcross progeny are reversed compared with those in the coupling backcross progeny; this is to be expected from the reversal of the arrangement of mutants in the double heterozygote. Thus wavy vestigial and normal mice are common in the coupling backcross progeny, rarer in the repulsion. Note also that mates suitable for a coupling backcross are again available, namely, the recombinants of the repulsion backcross (the two on the left).

References

Stern, C. (1960). *Principles of human genetics.* p. 449, W. H. Freeman.

Wallace, M. E. and Spickett, S. G. (1967). 'Hydronephrosis in mice, rats and man'. *J. Medical Genetics,* 4, 73–82.

7

Investigations with three linked loci

Linkage groups

In the last chapter it was shown how two linked mutants segregate through the generations, and that this depends on how they are arranged on the chromosomes of a homologous pair. There are some 200 mutants in the mouse whose positions on the 20 pairs of chromosomes in this species are known.

This 'mapping', as it is called, was done by making pairs of mutants segregate together until some combinations were found to behave in the manner described in the last chapter, rather than to segregate independently (as described in Chapter 5). Further mutants were then tested against these until, again, some were found to be linked with one or both members of a linked pair. Such sets of linked mutants were called 'linkage groups'.

Order

There then arose the question of order. Clearly if mutant t is closer to s than it is to r (all hypothetical), t is at the s end of the chromosome; but is it inside the segment r–s or outside it? If all three were closely linked, their recombination values would add, so that the value for the pair of mutants on the outside (say r and t) would equal the sum of the values for the two closer pairs (r,s and s,t); so, all one had to do was to decide which was the biggest value, and deduce that the pair of mutants showing this value was on a piece of chromosome with the third mutant inside.

Additivity

The reason for additivity is simply this: if segments r–s and s–t are very short, then a crossing-over event in r–s is rare, and is most unlikely to be accompanied by a similar event, also rare, in s–t. So by far the commonest crossing-over events will be (assuming for convenience the coupling phase for *r–s* and *s–t*):

Now, a cross-over between *r* and *s* is also a cross-over between *r* and *t*; similarly a cross-over between *s* and *t* is also a cross-over between *r* and *t*. So, since these happen on separate occasions, the sum of the *r–s* ones and the *s–t* ones is the total number of the *r–t* ones.

However, where three mutants were loosely linked, their recombination values were found to be not quite additive. One value was still bigger than the other two, and this was found to involve the outside pair as before; but non-additivity was found to depend on the very looseness of the linkages. This can be seen best if all three mutants *r*, *s*, and *t* are segregating at once. Then if crossing-over in both *r–s* and *s–t* is common, it will sometimes happen that crossing-over will occur in both at once:

Now, although two cross-over events have happened between *r* and *t*, the gametes' genotypes in respect of *r* and *t* will be just the same as if there had been no crossing-over at all, namely *rt* and ++. And this means, of course, that in a backcross progeny, the phenotypes will be non-recombinant instead of recombinant. The recombination value for the *r–t* segment now reflects fewer cross-over events than have really happened.

True, if all three mutants are segregating together, one could if one wished add in to the recombination value for *r–t* the events which, as seen from the progeny phenotypes, were double-cross-over ones. But this is rather pointless when one considers that for all loosely linked pairs of mutants where there are no extra mutants between them segregating, this adjustment is impossible. The function of recombination values, that of being a fairly rough measure of the distance between two mutants on the same chromosome, remains; one must simply remember that the further apart they are, the rougher is the measure. (There is also another phenomenon which affects additivity, but this will be examined later.)

Procedure and analysis

It remains to consider how best to investigate mapping procedures so that the above features will become apparent. Clearly, three fairly loosely linked mutants must be used, and they must be made to segregate simultaneously. As before, the backcross will be the most direct way of discerning through progeny phenotypes the gametic genotypes produced by the heterozygote (now to be a triple one); thence crossing-over events may be deduced. When the data become available they must be set out in an orderly and meaningful manner. A manner which allows easy calculation of recombination values follows.

Assume that r–s–t is the correct order. Assume also that the mutants are all in coupling (although heterozygotes with repulsion elements can be used); so, the linkage backcross is:

$$\frac{+++}{r\ s\ t} \times \frac{r\ s\ t}{r\ s\ t}.$$

The cross-over events will be as follows, taking the commonest first and the rarest last: no crossing-over at all, crossing-over in segment r–s and not in segment s–t, crossing-over in segment s–t and not in segment r–s, and crossing-over in segments r–s and s–t simultaneously. If segments r–s and s–t are labelled 1 and 2, then these events may be labelled in turn (o), (1), (2), and (1, 2). Event (o) will give gametes rst and $+++$, and each of the other events will similarly give two 'complementary' gametes (i.e. each gamete having what the other lacks). The threefold recessive will give only rst gametes, and these may for convenience be written on the lower line for each genotype of the offspring. The data may be then collected under the following heads:

	(0)		(1)		(2)		(1, 2)	
Progeny genotypes	$\frac{+++}{r\ s\ t}$,	$\frac{rst}{rst}$	$\frac{r++}{r\ s\ t}$,	$\frac{+st}{r\ s\ t}$	$\frac{++t}{r\ s\ t}$,	$\frac{rs+}{rst}$	$\frac{+s+}{r\ s\ t}$,	$\frac{r+t}{r\ s\ t}$
Progeny phenotypes	normal	rst	r	st	t	rs	s	rt
Observed numbers	large		medium		medium		small	

The size of the two medium pairs of numbers will depend on the sizes of segments r–s and s–t: if the former is the longer, the numbers under the left-hand 'medium' will be larger than those under the right-hand one.

The recombination values may now be obtained by gathering the observed numbers of recombinant phenotypes for each pair of mutants. Letting (0) now stand for the sum of the numbers of the two phenotypes under this heading, and so on for the other numbers, these values are:

For r, s: (1) + (1,2) divided by the total
For s, t: (2) + (1,2) divided by the total
For r, t: (1) + (2)   divided by the total

Note that the rarest event swaps the middle mutant and its allele compared with the arrangement in the triple heterozygote. Where order is not known, therefore, one can find it merely by seeing which pair of phenotypes is the rarest, and then seeing which pair of alleles has been swapped to produce it as compared with the arrangement in the triple heterozygote: the locus concerned with this pair is then in the middle. The event producing this pair is then (1,2), and the other labels can now be assigned. In other words, given the arrangement in the triple heterozygote but not the order, one need only list the phenotypes of progeny in pairs; then the observed numbers of these give the main clue to the order, the headings can be filled in, and the recombination values can be easily calculated.

Investigation 11: Three loosely linked loci

The choice of mutants depends not only on their linkage relations but on their viability and penetrance. The mutants tan-belly, a^t, wellhaarig *we*, and Danforth's short-tail *Sd*, are suitable in these respects (Plates 1 and 7). Classification should however be completed at the earliest possible age; if only one mutant is deficient in number due to early death, the recombination values will not be distorted, but if two are deficient, there will be distortion and the finer points of the investigation will be obscured.

These loci are in the order a^t–*we*–*Sd* and the triple recessive is non-agouti wellhaarig (or wavy). It is convenient, though not vital, to have other colour loci homozygous, e.g. that all the mice be brown, *bb*.

As the programme requires the rearing of a fairly large number of progeny in order that a meaningful analysis can be made, it is convenient to cut the number of generations to a minimum. Thus the first mating should be the coupling backcross: namely the mating of tan-belly short-tail mice of the following genotype to non-agouti wavy:

$$\frac{a^t + Sd}{a\ we\ +} \times \frac{a\ we\ +}{a\ we\ +}.$$

Note that the mutant symbols are not all on the same line as in the *r*, *s*, *t* description. This is because there has arisen in mouse genetics the convention of defining coupling as having the *dominants* on the same homologue, not the *mutants*. Where mutants are all recessive, this comes to the same thing. But where one is dominant, as in this case *Sd* is, the two definitions lead to different arrangements. If mutants were the criterion here, then the heterozygote above would have to have *Sd* and its + allele swapped. A heterozygote with *Sd* on the *a–we* chromosome would not be as convenient for the work, for it would not produce further triple heterozygotes and triple recessives except by crossing-over. The heterozygote as written above produces them both by the commonest event, no crossing-over. This means that further matings of the same type can be made easily from among the progeny.

The arrangement of the data for analysis corresponding to that for *r*, *s*, *t* is given below, and the headings assume that *a–we* is segment 1 and *we–Sd* segment 2. The observed numbers of the eight phenotypes of progeny from an actual experiment are also given.

(0)		(2)		(1)		(1,2)	
$\dfrac{a^t + Sd}{a\ we\ +}$,	$\dfrac{a\ we\ +}{a\ we\ +}$	$\dfrac{a\ we\ Sd}{a\ we\ +}$,	$\dfrac{a^t + +}{a\ we\ +}$	$\dfrac{a + Sd}{a\ we\ +}$,	$\dfrac{a^t\ we\ +}{a\ we\ +}$	$\dfrac{a^t\ we\ Sd}{a\ we\ +}$,	$\dfrac{a + +}{a\ we\ +}$
40	48	30	35	7	9	4	5
88		65		16		9	

E

Note that if the order were not known, it could easily be deduced from the observed numbers: 9 is the smallest for a pair of phenotypes, so these must have arisen by event (1,2); this event swaps the middle pair of alleles in respect of the arrangement in the triple heterozygote; *we* and its normal allele + have been swapped, so the *we* locus is in the middle. With the order a^t–*we*–*Sd* now established, the headings could also be filled in.

The recombination values estimated in the manner described above, for these data, are:

$$we\text{–}Sd: \quad 65 + 9, \text{ divided by } 178, = 42\%$$
$$a^t\text{–}we: \quad 16 + 9, \text{ divided by } 178, = 14\%$$
$$a^t\text{–}Sd: \quad 65 + 16, \text{ divided by } 178, = 46\%$$

Note that $42\% + 14\%$ is not 46%. Additivity could be achieved if the last estimate included $(2 \times 9)/178$. However, this is unrealistic, for *we* and *Sd* are so far apart that double crossing-over has probably happened between these two but cannot be detected by the progeny phenotypes; so this segment is probably longer in relation to the a^t–*we* than the estimated 42% makes apparent. If the *we*–*Sd* value cannot be adjusted, is there anything to gain by adjusting the a^t–*Sd*?

These data allow a rather more abstract phenomenon to be considered by those who are interested. It was stated earlier in this chapter that, if two adjacent segments are very short, the probability that crossing-over in one would be accompanied by crossing-over in the other is very small. It is in fact even smaller in mice (and in many other organisms) than one might suppose from considering their rarity alone; for the chances of simultaneous crossing-over in adjacent segments are not independent. The occurrence of one cross-over inhibits the occurrence of another close to it, i.e. one event interferes with the other. This phenomenon is in fact called interference. To show this, the above data may be arranged as follows:

		Segment *a–we*	
		Non-recombinants	Recombinants
Segment	Non-recombinants	88	16
we–Sd	Recombinants	65	9

It is now apparent that recombination in the *we–Sd* segment is smaller when there is also recombination in the *a–we* ($9/(9 + 16)$ = 36%) than when there is no recombination in the *a–we* ($65/(65 + 88) = 42.5\%$).

There are at present no stocks commercially available for this type of investigation, other than that described above containing a^t, *we*, *Sd*, and *b*, although others may become available in the future.

Investigation 12: Three loosely linked loci and one independent one

The data given in the last section were obtained from an experiment in which brown, *b* (Frontispiece), and its black allele, $+$, were also segregating. It was done to check that the brown locus is not linked with any of the others. It was found as easy to classify a^t, *we*, and *Sd* in black mice as in brown, so the stock made available commercially contains these two loci segregating also, for those wishing to make the same check.

In order that putative linkage data for $+$, *b* be provided by the backcross, the triple heterozygote for a^t, *we*, and *Sd* must clearly be heterozygous for *b* also; the desired mating may be written thus:

$$\frac{a^t + Sd +}{a \ we + b} \times \frac{a \ we + b.}{a \ we + b}$$

The writing of $+$, the normal allele of *b*, on the same line as $a^t + Sd$ indicates that black came into the heterozygote from the same parent as a^t, $+$, and *Sd*; if there is linkage it is thus in coupling with each of the genes. The break in the line dividing the two halves of the heterozygotes' genotype indicates that linkage is uncertain (as it is to be tested), so there may be two pairs of homologous chromosomes involved here.

There are several ways of setting out the data for analysis. A simple one is to take each pair of possible linkages, a^t–*b*, *we*–*b*, and *Sd*–*b* in turn and write out for each, first the non-recombinant pair of phenotypes and their observed numbers, and then the recombinant pair and their observed numbers, as for the *wa*–*vt* coupling backcross in Chapter 6. This gives three small tables of data, the progeny total being the same for each one. This can be done by the reader from the data in the last section divided into black and brown progeny. The 40 progeny on the left of the table in that section consisted of 21 black and 19 brown; the 48 progeny next on the right were 24 blacks and 24 browns, and so on. Further divided in this way, the data are:

Black:	21	24	12	20	4	6	3	0	90
Brown:	19	24	18	15	3	3	1	5	88
	88		65		16		9		178

Continuance of this stock is similar to that described in the last section. Progeny of the same genotypes as the parental back-cross mating should be chosen and mated together. They will occur about half as frequently as before (the fourfold heterozygote in the data above occurs 21/178 progeny, and the fourfold recessive 24/178 progeny); for this reason it is wise to start with several backcross matings, and to mate suitable progeny as soon as they have been classified and are weaned.

Reference

Wallace, M. E. (1957). 'A balanced three-point experiment for linkage group V of the house mouse'. *Heredity*, 11, 223–58.

Glossary

(Abbreviated, with some rephrasing, from *The Elements of Genetics*, by C. D. Darlington & K. Mather, 1949. Allen & Unwin.)

Allele: one of two or more dissimilar genes having the same position (locus) in one of a pair of homologous chromosomes. (Same as allelomorph.)

Autosome: a chromosome whose segregation from its partner at meiosis does not affect the determination of sex.

Backcross: the cross of a hybrid with one of its parents, or a genetically equivalent cross.

Breed: a variety, in domestic animals.

Character: a property of an organism in regard to which genetic similarities or differences of individuals are recorded.

Chiasma (plural **chiasmata**): an exchange of partners in a system of paired chromatids, observed during meiosis.

Chromatid: a half-chromosome formed by the splitting of a chromosome during meiosis into two similar halves.

Chromosome: one of a group of threadlike bodies in the cell which carry the genes or units of heredity.

Chromosome map: a diagram showing the linear order of the genes or of gene-positions within the chromosome.

Coupling: the presence of two given genes in the same chromosome in a double heterozygote, as opposed to Repulsion, where they are in different homologous chromosomes.

Cross: an act or product of cross-fertilization.

Crossbreeding: outbreeding.

Crossing-over: the exchange of corresponding segments between chromatids of homologous chromosomes, by breakage and re-union following pairing: a process inferred genetically from the recombination of linked genes in the progeny of heterozygotes, and cytologically from the formation of chiasmata between homologous chromosomes.

Cross-over value: the frequency of crossing-over between two genes or markers. Often loosely used for Recombination Value to which it may be equated only when small.

Dihybrid: heterozygote in respect of genes at two different loci.

Diploid: an organism having two sets of chromosomes, as opposed to organisms having one (haploid), three, or more sets. Also, the zygotic number of chromosomes ($2n$) as opposed to the gametic or haploid number (n).

Dominance: the relationship of two alleles where the single gene heterozygote resembles one of the two homozygous parents (said to carry the dominant allele) rather than the other (said to carry the recessive allele) on an arbitrary scale distinguishing the two phenotypes. Partial dominance = incomplete dominance; some would call this lack of dominance.

Dominant mutant: term applied in *Drosophila* to any mutant whose effect is detectable when heterozygous with its wild-type allele.

Double heterozygote: heterozygote in respect of genes at two different loci.

Expressivity (of a gene): the degree of manifestation of a genetic effect in those individuals in which it is detectable. Expression: the manner or description of manifestation.

F1: the first generation of the cross between two individuals homozygous for the particular genes which distinguish them.

F2: the second filial generation obtained by self-fertilizing or crossing *inter se* individuals of an F1.

Gamete: cell which is specialized for fertilization; egg or sperm. The gametes of a diploid organism carry one set of chromosomes (n).

Gene: any particle to which the properties of a Mendelian factor may be attributed.

Genetic: pertaining to or analogous with heredity.

Genotype: the kind or type of the hereditary properties of an individual (*see* Phenotype).

Heterozygote: a zygote derived from the union of gametes dissimilar in respect of the quality, quantity, or arrangement of their genes. Usually used in respect of particular gene differences.

Homologous chromosomes: chromosomes of similar structure which pair at meiosis. In diploids, the two sets of chromosomes, one set from one parent and the other from the other parent, come together in pairs (n pairs) during meiosis.

Homozygote: a zygote derived from the union of gametes identical in respect of the quality, quantity, and arrangement of their genes, or of certain of them.

Inbreeding: the raising of progeny by the mating of two more closely related zygotes, as opposed to outbreeding from the mating of two less related zygotes.

Interference: the property by which one cross-over interferes with the occurrence of another cross-over in its neighbourhood.

Karyotype: the character of a nucleus as defined by the size, shape, and number of the mitotic chromosomes.

Linkage: the combination in the gametes formed at meiosis of pairs of segregating genes in non-random frequencies owing to their presence in the same chromosome or in two chromosomes paired at meiosis.

Locus: the position occupied by a gene in a chromosome, with regard to its linear order.

Meiosis: a double mitosis in which the nucleus divides twice but the chromosomes only once. It occurs prior to gamete-formation.

Mendelian inheritance: obeying Mendel's laws. Particulate as opposed to blending.

Mendel's laws of inheritance: Law 1: the law of segregation, that the gametes produced by a hybrid or heterozygote contain unchanged either one or the other of any two factors determining alternative unit characters in respect of which its parental gametes differed.

Law 2: the law of recombination, that the factors determining different unit characters are recombined at random in the gametes of an individual heterozygous in respect to these factors. (Note that linkage is an exception to this law.)

Mitosis: the process by which division of the nucleus is accomplished by that of its constituent chromosomes and usually accompanied by that of its containing cell. It occurs during growth of the body of an individual.

Monohybrid: heterozygote in respect of one gene.

Mutant: aberrant individual, cell, or gene produced by mutation.

Mutation: a change of heredity not ascribable to segregation or recombination, e.g. from one gene to an allele.

P: the parental generation of an F1.

Pedigree: table of ancestry or of posterity.

Penetrance (of a gene): the proportion of individuals of a given genetical constitution in a given population in which the effect of the gene concerned phenotypically distinguishes them from those bearing its allele.

Phenotype: the kind or type of organism produced by the reaction of a given genotype with a given environment.

Pleiotropy: the production of physiologically uncorrelated effects by a mechanically unitary, i.e. single, gene. Ascribed to one gene initiating two or more chains of reactions.

Population: a mating group limited for special consideration by either environment or breeding system.

Progeny test: the method of assessing the genetic character of an individual by the performance of its progeny.

Recessive: *see* Dominance.

Recombination: the formation by crossing-over or segregation at meiosis of new combinations of genes with respect to (i) individual chromosomes or (ii) whole gametes.

Repulsion: *see* Coupling.

Segment: a portion of a chromosome considered as a unit for a given purpose.

Segregation: separation at meiosis of the chromosomes, or parts of chromosomes such as genes, of paternal and maternal origin.

Selective advantage: that genotypic condition of an individual or genetic class of individuals which increases its chances, relative to others, of representation in later generations of individuals.

Sex chromosome: one whose distribution to one and not to another of the products of meiosis determines the difference in sex of the offspring.

Sex-limited: of the inheritance of differences which are expressed in one sex only, or at least in the two sexes differently.

Sibs (or siblings): progeny of the same parents derived from different eggs. Sib-mating is the mating of sibs, and is the closest form of inbreeding in diploids where each individual can be of only one of the two sexes.

Stock: an artificial mating group, usually for the preservation of certain genes or combinations of genes.

Strain: a natural or artificial mating group uniform in some particular. An inbred strain is one produced by the mating of relatives. A standard inbred strain of mice is one produced by the mating of sibs for twenty generations.

Testcross: a cross of a double or multiple heterozygote to the corresponding double or multiple recessive. Used to estimate linkage relationships or behaviour.

Wild type: of an organism or gene of the type predominating in the wild population.

X chromosome: with diploid sex differentiation, the sex chromosome in regard to which one sex is homozygous.

Y chromosome: with diploid sex differentiation, the sex chromosome that is present and pairs with the X chromosome in the sex which is heterozygous.

Zygote: cell formed by the fusion of gametes, and the individual derived from it.

AN LEABHARLANN,
CEARD CHOLÁISTE RÉIGIÚN
LEITIR CEANAINN.

Index